PROGRESS of the BREED

PROGRESS of the BREED

The History of U.S. Holsteins

Richard H. Mansfield

Centennial Edition

Edited by Robert H. Hastings

Holstein—Friesian World, Inc.

International Standard Book Number: 0-9614711-0-7
Library of Congress Catalog Card Number: 85-060730
Holstein-Friesian World, Inc., Sandy Creek, NY 13145

Published 1985
Printed in the United States of America

This book is dedicated in memory of
Helen Prescott Hastings and Jean Montgomery Rumler.

Contents

Foreword **XII**

Preface **XIX**

PART 1 THE HOLSTEIN-FRIESIAN ASSOCIATION

Introduction **3**

1. **THE EARLY YEARS** **5**
   - Early Imports 6
   - Rival Breed Associations Are Formed 7
   - Holstein-Friesian Association of America 8
   - Early Recognition For Holsteins 10
   - Houghton Takes Over 12
   - Organizational Changes 15
   - Production Testing Begins 15
   - Building A Field Force - The Beginning 17
   - Holstein Youth Programs 20
   - Maintaining Integrity - The Cabana Case 22
   - The Proxy Fight 23
   - True Type 25

2. **THE DEPRESSION AND WAR YEARS** **29**
   - Complete Herd Testing (HIR) 30
   - Herd Classification 31
   - Sire Recognition 33
   - Progressive Breeders Registry 34
   - A Permanent Home For The Association 35
   - The Association Weathers The Storm 35
   - Wartime Growth 37
   - The Field Force Expands 38
   - Problems At The Office 40
   - The Efficiency Study 41
   - The Association Reorganizes 43

3. **GETTING DOWN TO BUSINESS** **45**
Changes at the Home Office 45
The Phenomena of AI 47
Keeping Records Credible With Blood Typing 49
Parallel Breed Improvement 50
State And Local Holstein Clubs 52
The Field Force Builds 53
Recognizing Brood Cows 55
Finishing The Fabulous Fifties 56

4. **BROADENING THE GENETIC BASE** **59**
The Computer Age Arrives 61
Maintaining Integrity 63
The Harrisburg Association Disbands 63
Struggle Against Traditionalism - The Color Controversy 64
Advances In Evaluating Type - Descriptive Classification 67
The Field Force - A Period Of Transition 71
Developments In Production Testing 72
New Breed Improvement Tools - Sire Summaries And Pedigrees 74
Genetic Advances In The Late Sixties 76

5. **REACHING OUT TO THE WORLD** **79**
Broadening The Horizons To The World Community 80
USDA/FAS Cooperator Program 82
International Export Development 83
Holstein-Friesian Services, Inc. 85
International Achievements 93

6. **A FULL SERVICE ORGANIZATION** **99**
Genetic Evaluation And Management Service 100
Holstein Sire Development Service 102
State Organizations - Termination of the Allotment System 104
Improving Classification Data 106
More Tools For Genetic Advancement 107
Sire Evaluation For Type (SET) 109
New "Ideal" Type Models 109
Identified Holstein Female - Grade I.D. Program 111
The Rearick Case 112
Undesirable Recessives 113
Accomplishments And Changes 116

7. **A GLOBAL BREED ORGANIZATION 119**
Management Initiatives For The 1980's 119
Linear Classification 125
Embryo Transfers 126
World Friesian Conferences 128
International Ethics 128
Success In Hungary - An Example And A Model 129
A Century Of Progress 132

PART 2 HOLSTEIN ACTIVITIES

Introduction 157

8. **PUBLIC SALES 159**
Promotional Sales 163
Dispersal Sales 164
Early Sales of Holsteins 165
Higher Production - Higher Prices 166
The Cabana Aftermath 170
The Great Depression 174
The Sales Boom of World War II 175
The Super Duper Decade 178
Soaring Sales 181
A "Bull Market" For Holsteins 186
New Times - New Ideas 190

9. **THE SHOW RING 197**
The Early Years 198
The National Shows Before World War II 200
The Dairy Cattle Congress 203
International Dairy Show 204
The Three National Show Concept 206
Production Recognition 209
The Three National Shows 211
State Association Shows 215
The Role of the Holstein Association 217
The All-American Program 218
An All-American Update 221
All-Time All-Americans 225
Accomplishments of the Show Ring 230

10. ORGANIZATIONS THAT HELPED BUILD THE HOLSTEIN BREED 235
The Breed Publications 235
Purebred Dairy Cattle Association 244
The Dairy Shrine 246

PART 3 HOLSTEIN SIRES OF PROMINENCE

Introduction 253

11. FAMOUS PRE-AI SIRES 255
Johanna Rag Apple Pabst 256
Governor Of Carnation 259
Dunloggin Woodmaster 263
Wisconsin Admiral Burke Lad 266

12. PROMINENT SIRES OF THE EARLY AI YEARS 271
ABC Reflection Sovereign 271
Wis Burke Ideal 274
Pabst Sir Roburke Rag Apple 276
Burkgov Inka De Kol 278
Searsfarm Dean Ada Imperial 282
Osborndale Ivanhoe 285
Romandale Reflection Marquis 289

13. GREAT SIRES OF THE "GENETIC ERA" 293
Whirlhill Kingpin 294
Ideal Fury Reflector 298
No-Na-Me Fond Matt 301
Pawnee Farm Arlinda Chief 306
Penstate Ivanhoe Star 309
Paclamar Bootmaker 313
Paclamar Astronaut 316
Round Oak Rag Apple Elevation 320
Glendell Arlinda Chief 324

APPENDIX 329
GLOSSARY 351
INDEX 355

Foreword

One hundred years have passed. Now, it's 1985! Where is the Holstein breed and Holstein-Friesian Association of America today?

Like a personal biography, the history of an organization is of little importance unless its achievements are noteworthy, its contributions significant and its stature monumental. Holstein-Friesian Association of America qualifies as does the Holstein breed in the United States.

In its first 100 years of existence, Holstein-Friesian Association of America has risen to an undisputed position as a premier herd book organization. But this status of leadership comes not alone from its activities as a registry organization for Holstein-Friesian (now HOLSTEIN) dairy cattle in the country. It comes also from its well-rounded, complete program of activities, quality services and high standards which it provides the Holstein industry in the United States and, to a measured degree, worldwide. This "full-service" structure makes it unique among its peers.

Few organizations devoted to livestock interests have achieved the sustained forward movement of the Association over such a long period of time.

Through the years, the Association and the breed faced matching challenges and opportunities but none equaled the last three decades. It was during this time that the breeding of dairy cattle made the transition from an art to a science — that technological advances were readily introduced and implemented — that reliable genetic data and information became critical to improvement of the species — that industry integrity and credibility became an unqualified, fundamental hallmark for the Association to be able to serve its true purpose. It was also during this time that flexibility and adaptability, advanced planning, a progressive philosophy, operational management based on sound business principles and qualified leaders became key elements in the Association's consistent rise to its premier status.

In marking the last three decades of history as a period of great expansion and advancement to a status of stardom, the heritage and solid foundation of the preceding 70 years played its role, although the position of the Association was not as spectacularly dominant in the industry as it is today.

Restructuring Management

The dynamic, post World War II growth of the Association to its premier position was triggered by a restructuring of the organization in 1948. A major rewriting of the bylaws abolished the long-standing management of the operational sectors by committees of the Board of Directors. This antiquated system was replaced by executive management. The Executive Secretary became the corporate equivalent of the Chief Executive Officer with matching responsibilities and authority. The Board of Directors assumed its present posture in policy-making while retaining "control and management of the affairs and business of the Association."

Forward Perspective

The single most significant factor in the rise of the breed and the Association to its present position was the perception of its leadership of what lay ahead. A carefully executed program of planning was the result, translating projected developments in the industry environment into policies and related courses of action for the Association.

Three decades of post World War II planning embraced the philosophy that "continued success depends on the skillful wedding of long-range planning [then in increments of 10 years each], annual program development and day-to-day execution." It was "a strategic approach to progress." It proved to be so.

More recently, planning has been updated to shorter spans of five years, reflecting the quickening pace of industry advances.

Now It's Holstein — Again

As detailed later, the first herd book of the breed in the United States, The Association of Breeders of Thoroughbred Holstein Cattle, was organized in 1871. It was the first herd book of any breed or species of livestock in the world. The breed in the U.S. was officially designated as HOLSTEIN. Later, in 1877, a rival herd book was organized as the Dutch-Friesian Breeders' Association, but it was the Holstein herd book which dominated the recording process in numbers and membership.

When, in 1885, the two organizations wisely decided to join forces, one of the compromises was the selection of key words from the name of each organization in establishing the official name of the new herd book association. And so, "Holstein" and "Friesian" were blended into "Holstein-Friesian" in the new organization's corporate name and the official name of the breed in the U.S.

The name prevailed until 1977, when a pronouncement was made to the annual meeting of Association delegates that the time had come to clearly define and designate through its name the unique characteristics of the breed as developed by more than 70 years of highly selected, single-purpose, closed herd book breeding by U.S. breeders in accordance with guidelines and improvement programs set forth by Holstein-Friesian Association of America. So, once again, it became the HOLSTEIN breed in the U.S., but the similarity with the original remains only in name and ancestry, for the genetic pool of the breed in the closed herd book of the Association today represents the world's largest, purest, and most genetically advanced.

The long-term significance of the name change is gradually becoming evident in the dairy community of the world as a greater and greater volume of U.S. genetic material is being selected and purchased for international use, especially by the developed dairy countries of the world. The mark of "Holstein" breeding will be etched clearly on the industry worldwide in the future.

World Involvement

Until 1905, the U.S. dairy industry was an importer of Friesian breeding stock from Europe. Thereafter, there were no imports and the Association operated, as it continues to do today, a closed herd book — which it considers to be fundamental to its role in serving the best interests of the breed and breeders, domestically and worldwide.

Perhaps one of the great stories of international livestock breeding is found in the "reverse osmosis" within the "Holstein" and "Friesian" sectors of the breed. Nineteenth century Friesian imports from the European continent, bred and highly selected for upwards of 80 years in the United States, now provide the genetic pool for selected breeding stock returning to the Continent and elsewhere in the world as Holsteins.

Domestic Position

An estimated 90 percent of the national herd of dairy cattle in the United States is made up of Holsteins. This dominance is the result of consistent advancement since World War II when it was projected at less than 50 percent. The ratio of herd book (registered) Holsteins in the national population of all registered dairy cattle stands at 80 percent as of this centennial year. The postwar figure was 46 percent. Registration of Holstein animals in the herd book of the Association is at the annual rate of one animal for every 24 cows in milk (all breeds) in the country.

Despite its closed herd book status, the Association does not preclude the identification and recording of Holstein females not otherwise eligible for entry in the herd book. Identification with a record of ancestry and related detail is available and in use as the "Identified Holstein Female" program.

The Association's International Herd Book serves to register animals and their offspring which may be imported and which were previously registered in the recognized herd book of another country. While participation is limited, it is a valuable and necessary adjunct to the recording program, domestically and internationally.

Genetic Improvement And Progress

The rate of genetic improvement in the breed must be considered phenomenal. For the registered population of Holstein cattle, the annual rate of genetic improvement is 150 pounds of milk and five pounds of butterfat. While conformation traits (type) are more subjective, genetic advancement is also at a remarkable level. The graphic story is told in the following three charts.

Genetic Trend For Milk Production

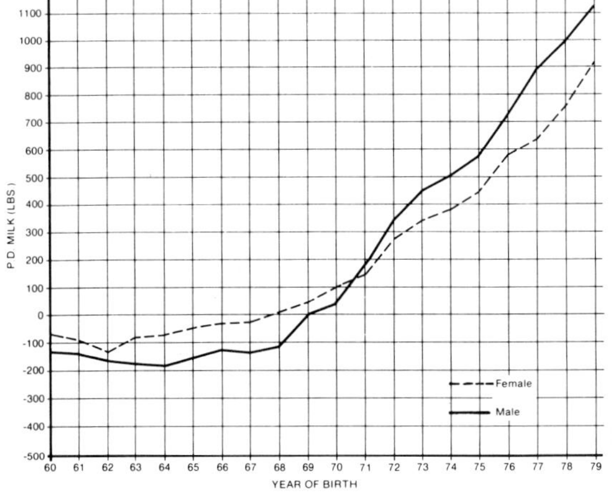

Genetic Trend For Fat Production

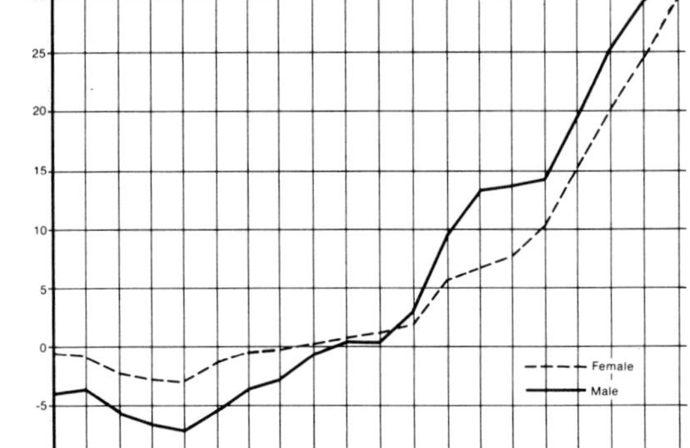

Genetic Trend For Conformation

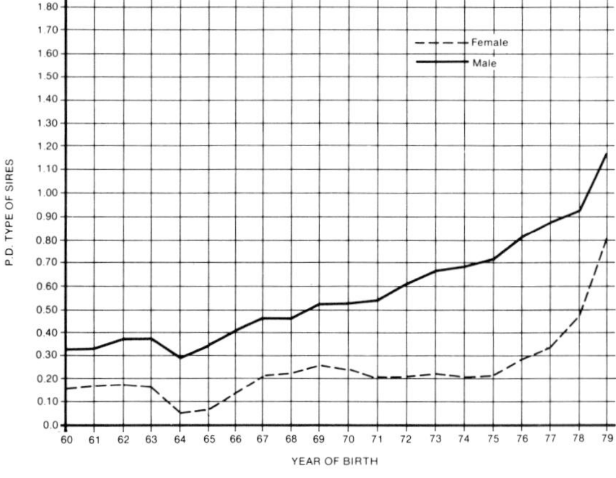

Heavy participation in programs of production testing (milk recording), classification (evaluation for conformation) and animal identification coupled with readily available, highly reliable, genetically-based breeding information on both sires and dams using state-of-the-art computer capability, consistently maintained by the Association as a matter of policy since 1950, makes this possible.

Fundamental to this success is the fact that the Association maintains the only complete file of genetic data on individual Holstein animals in the country. The primary reference, known and accepted worldwide, is the automatically produced, computer-printed, official Association pedigrees of Holstein animals. Annual demand for these pedigrees exceeds 200,000. Equal in stature, worldwide, is the semiannual "Sire Performance Summary," a complete reference to the transmitting pattern of Holstein sires as measured genetically through the performance of their offspring.

The preceding paragraphs give the reader a running start on the history which follows ... a preview of where the Association stands today and how the breed has progressed in the hands of the skillful breeders in the United States. The blending of the capabilities of the breeders and the services of the Association to them and on their behalf will become self-evident. This fascinating story is one of the truly great sagas of American agricultural enterprise. The breed of cattle, the breeders, and their organization have been and are an inseparable combination in producing what is recognized as the current model to emulate.

It took a century to happen.

Robert H. Rumler
Chairman Emeritus

Holstein-Friesian
Association of America

Preface

This book is the result of a team effort. At *Holstein World,* many of the staff and personnel handled research, editing, and production in addition to their regular duties. Notable among them are Jean Hager, Dawn Lindsey, Barb Clark and Cindy Bouchard, along with production team members Bill Woodruff and Diane Moffet.

Zane Akins and his Holstein Association staff gave explanations and helpful advice throughout the project. Special thanks are due to Ben Coplan for his energetic support and to Mona O'Connor for her thorough research. Jim Yanizyn did an excellent job of locating many of the photographs used in the book.

Major writing contributions were made by Jim Hill (All-Americans) and Ron Buffington (Prominent Sires). Others who contributed material were Ted Prescott (Breed Publications), Bradley Rugg (World Dairy Expo), and Jim Leuenberger (Dairy Shrine).

Horace Backus helped with the material on public auction sales and provided thoughtful review comments on the book. Jim Hill, Ted Prescott and Ron Buffington also helped with their review of the text.

Bob Rumler gave freely of his time and talents to the project in numerous interviews. His sense of history and his pride in the achievements of Holstein breeders brought a special dimension to the book.

My wife, Mary Robb Mansfield, deftly used her editorial skills on muddy writing, sexist language, and jargon. Her support and guidance were invaluable throughout the project.

This book would not have been completed without the work of Bob Hastings. From initial design to final picture selection, his editorial expertise has guided the project. The countless hours that he devoted to the book are reflected throughout the pages that follow.

RICHARD H. MANSFIELD

PART 1

The Holstein-Friesian Association of America

One Hundred Years of Progress

Introduction

Nestled into a rocky hillside overlooking the Connecticut River, the Holstein Association office anchors the southern end of Main Street in Brattleboro, Vermont. Most people in town are unaware that each day in this brick-faced building, hundreds of Holstein dairy cattle are registered, transferred, even bought and sold. Furthermore, few townspeople know that this corporation, where their friends and neighbors work, houses the genetic lifeblood of the whole U. S. Holstein industry and that on computer tapes whirring there, animals can be traced back to their European ancestors.

In 1985, the Holstein Association observes 100 years of progress — a centennial celebration that is well earned. This history tells of how a dynamic organization was built, how the integrity of the vital pedigree records has been maintained, and how, by mixing financial stability with forward-looking leadership, the Holstein Association has reached its present position of prominence in the dairy industry of the world.

Growing from a small group of 284 breeders at the start to over 44,000 members today, the Association has led the growth of the Holstein breed from a handful of imports to a position of leadership in the international dairy industry. U. S. Holsteins are found in virtually every milk-producing country in the world and produce 90 percent of the milk consumed in the United States. And as they head into their second century, the Holstein Association staff and membership like it just that way.

Here is their history.

U.S. Holsteins

Holstein Association, Brattleboro, Vt., USA

chapter 1

The Early Years

The Holstein-Friesian Association of America was organized May 25, 1885 for "the purpose of improving the breed of Holstein-Friesian cattle, ascertaining, preserving and disseminating, as provided by its bylaws, all useful information and facts as to their pedigrees and desirable qualities, and the distinguishing characteristics of the best specimens, and preparing, publishing and supplying all necessary volumes of the Holstein-Friesian Herd Books; and, generally, for promoting and securing the best interests of the importers, breeders and owners of said cattle, and thereby the public generally."

Winthrop W. Chenery was the first to recognize the potential of Holsteins.

It was a tall order for the 66 breeders present at the organizational meeting to fill, but the goals set that day in Buffalo, NY, have served the Holstein breed well. The Association has grown from 284 members in 1885 to over 44,000 in 1985.

Early Imports

The first importations of Holstein cows to America were made by the early Dutch settlers in New York (New Amsterdam) in 1613 and later by the West India Company in about 1625. Later, in 1795, the Holland Land Company sent eight animals to the agent, John Lincklaen. The animals were reportedly described in this way: "The cows were the size of oxen, their colors were black and white, not spotted, but, large patches of the two colors; very handsome bodies and straight limbs; horns middling in size, but gracefully set; their necks were seemingly too slender to carry their heads; their disposition mild and docile."

In 1810, William Jarvis of Weathersfield, VT, a noted importer of Merino sheep, imported a bull and two cows. Shortly afterwards, several others, including Herman LeRoy of New York, brought Dutch cattle to the United States. The ancestry of all these early animals was lost through crossbreeding.

The real beginnings of the breed trace to the efforts of Winthrop W. Chenery of Belmont, MA, who, in 1852, purchased a Dutch cow from the master of a sailing vessel that had just landed a cargo of Holland rum at Boston. Chenery, impressed with the milk production of the animal, imported a bull and two cows in 1857 and four more cows in 1859. These animals contracted a pneumonia-type illness (Rinderpest), and all except a young bull, Dutchman 37, were seized and destroyed by the Commonwealth of Massachusetts as part of its effort to stamp out the disease. But this didn't stop Chenery, who had been so pleased with the Holsteins that he imported another bull and four more cows in 1861. These animals, while having a limited impact on the breed due to their small numbers, spread the fame of the Dutch cow across the country. Soon other dairymen, including Gerrit S. Miller of Peterboro, NY, started importing high producing Holstein dairy cattle. By 1872, animals with lineage from the Chenery stock had spread to 12 states and as far west as California.

According to Professor Henry Hiram Wing of Cornell University, who was an

Gerrit Miller, premier importer and breeder of Holsteins in the early years.

authority on the origins of the Holstein breed, probably the most important early purchases of Dutch cattle were those made by Gerrit Miller, who in 1869 imported a bull, Hollander, and the cows, Crown Princess, Dowager, and Fraulein. Miller, who later imported the famous sire Billy Boelyn and two renowned cows, Empress and Ondine, bred Holsteins in his Kriemhild herd from 1869 until the mid-1930's, raising animals which were outstanding producers for their time. Their progeny formed the foundation of many of the early Holstein-Friesian herds. (Ed. note: A definitive report of the origins of the Holstein breed by Professor Wing is contained in the 1930 and 1960 editions of the **Holstein-Friesian History**)

Miller, in a 1923 letter to Frank N. Decker, described the first milk record set in the United States, 12,681 pounds by Dowager No. 7, a record that did much to spread the fame of the Holstein breed. Miller wrote, " In 1869 a cow that would give 6000 lbs. of milk in a year and 12 lbs. of butter in seven days, was considered exceptional — Dowager was milked only twice a day, she had no grain during June, July and August, the remainder of the year she had from four to eight quarts of grain — half of it wheat bran each day — 50 lbs. per day was her largest yield which she gave soon after freshening and again six months later. She was a good cow to maintain her flow of milk."

Rival Breed Associations Are Formed

By 1871, there were enough breeders interested to form an Association to record the pedigrees of the imported animals and their offspring. On March 15, 1871, The Association of Breeders of Thoroughbred Holstein Cattle was formed in Boston. Mr. Chenery was its first president and in 1872, he published its first herd book which listed 128 animals. An energetic writer and advertiser, Chenery was instrumental in bringing the milk-producing qualities of the Holstein breed before the American public. He also sold foundation animals to many who would go on to become prominent Holstein breeders, including Charles Houghton, Thomas B. Wales, William A. Russell, George E. Brown, and Dexter Severy. He continued to lead the organization until his death in 1877.

During this formative period, Thomas E. Whiting of Concord, MA, began

Solomon Hoxie led the Dutch Friesian Association of America. He later brought Advanced Registry to the Holstein-Friesian Association of America.

importing Dutch cattle. The subsequent feud between Chenery and Whiting is an interesting part of the early history of the Holstein-Friesian breed. Whiting, who has been credited with encouraging the formation of the Netherlands Herd Book Association, imported 28 animals between 1871 and 1875. For reasons unclear from past writings, he could not obtain registration for all his imports so he started his own rival herd book, tracking his animals and their offspring. The herd book was in manuscript form at his death.

During the same period, a group of young farmers in Otsego County, New York, led by the spirited Solomon Hoxie, formed a business organization called the Unadilla Valley Stock Breeders Association and in 1874 and 1876, purchased animals from Thomas Whiting. After Whiting's death in the summer of 1877, the group bought the rest of his herd as well as the herd book manuscript and immediately formed a herd book association called the Dutch Friesian Association of America. Led by Hoxie, the members of this new group had some strong feelings about imports. They felt that animals should be imported only from certain areas of Europe, in contrast to the more liberal interpretation used by the Holstein organization. Dutch Friesian members also believed that the name "Holstein" was incorrect for their Dutch animals and developed the slogan: "A Pure Register and a Correct Name." The Dutch Friesian group also developed some important selection and testing concepts that later, after the merger of the two groups, would lead to the Advanced Registry and Classification programs. But before that would happen, a rivalry, bitter at times, persisted between the two groups from 1877 to 1885 with acrimonious articles and stiff rebuttals flooding the agricultural press. Soon, economic factors would end the competition and arguments, bringing the two battling organizations toward union.

Holstein-Friesian Association Of America

The years from 1877 to 1885 marked the permanent establishment of Holstein cattle in the United States. The Prairie States were filling with settlers; new inventions in agricultural machinery sparked an agricultural boom. Starting with the Chenery importations, about 6500 animals had been imported from Holland when an outbreak of hoof and mouth disease in Europe virtually dried up the foreign source of cattle. Because of this crisis, American breeders decided that it was time to stop feuding and get down to the business of protecting and promoting the Holstein breed. And just in time, because imports dropped drastically in 1885 and stopped altogether in 1905. There have been no importations from outside North America since.

Year	Number Imported	Year	Number Imported
1852	1	1880	400
1857	3	1881	229
1859	4	1882	173
1861	5	1883	1059
1865	1	1884	1515
1869	12	1885	2538
1871	8	1886	526
1872	3	1887	281
1874	97	1888	76
1875	8	1889	56
1876	40	1890	103
1877	31	1903	13
1878	23	1904	89
1879	433	1905	30

By 1885, the differences between the two rival breed organizations had been all but resolved, except for the name "Holstein." But, there were other factors of concern. The highly publicized records made by several leading Holsteins had stimulated an active demand. Before the hoof and mouth outbreak, importers had flourished during 1882, 1883, and 1884. However, the large importations of animals had flooded the market with, in some cases, poor quality animals. At the same time, some sharp Hollanders were bringing in cattle to the U.S. to sell to dairymen and returning home with dollars in their pockets, dollars that American importers wanted. A united front was needed. At the annual meetings of both organizations in the spring of 1885, a joint committee was set up to join the two organizations together.

On May 25, 1885, two months after Grover Cleveland's first inaugural, the Holstein-Friesian Association of America was formed. The 284 charter members

Thomas B. Wales, Jr. was the first Secretary of the Holstein-Friesian Association of America.

came from 26 of the 38 states then in the Union.

The original charter called for a government by a Board of Officers, consisting of a president, four vice-presidents, six directors, a secretary-treasurer, and a superintendent of Advanced Registry. There was a harmonious division of offices between the two former rival groups. Thomas B. Wales, Jr., who had purchased his first animals from Mr. Chenery and had imported 300 Holsteins in 1884, was elected Secretary. Wales had been an active participant in the Dutch Friesian battles and had often fenced with Solomon Hoxie in the agricultural press. Hoxie, who had been Secretary of the Dutch Friesian group, was elected Superintendent of Advanced Registry. Theron G. Yeomans, head of the Holstein group was elected President, while the President of the Dutch Friesian organization, F. W. Patterson, was made the First Vice-President. The personal antagonisms of the past faded as the newly-organized group took up the task outlined by their predecessors: "improving the breed of Holstein-Friesian cattle...."

The first years of life of the new Association were rocky. Sales of Holstein cattle plummeted. Breeders were disenchanted with the breed: the market had been glutted by an oversupply of cattle from earlier importations, many of which were of poor quality and did not produce at the level dairymen had anticipated. The crowning blow, in an era of butter and cream, came from competition from other breeds. In the days before the centrifugal separator, it was difficult to separate the cream from Holstein milk — Jersey and Guernsey milk made butter more easily and the already well-established Channel Islands breeds found favor with many dairymen.

Early Recognition For Holsteins

In May, 1887, a noteworthy event in the history of the Holstein breed in America took place. It was the Madison Square Garden dairy cattle show where the four leading dairy breeds, Ayrshire, Jersey, Guernsey, and Holstein-Friesian, met for the first time to see which was the greatest producer of milk and butter. Prizes of $200 were offered for both 24-hour milk production as well as butter production. Most observers conceded that a Holstein would win a milk

Many were surprised when Clothilde won this silver cup for butter production in 1887.

Early butter test winners

Clothilde

Mercedes

production prize, but the Jersey breeders were certain that they would take the butter prize — so certain that they offered a handsome silver cup, with a beautiful Jersey cow engraved on the side, to the winner. But that cup is now sitting in the Holstein Association office in Brattleboro, VT. When the butter samples were weighed, Clothilde, a Holstein owned by Smiths & Powell of Syracuse, had won the $200 and the silver cup.

This decisive victory in a public butter test, which followed on the heels of the triumph in 1883 by Thomas Wales' imported cow, Mercedes, over the famous Jersey cow, Mary Ann of St. Lamberts, caught the attention of dairymen across the nation at a time when butter production was important and all dairy breeds were fighting for recognition. Holstein breeders were quick to follow up on these trail-blazing successes, and the consistent victories in competition played a big part, especially in the Midwest, in the rapid expansion and popularizing of the Holstein breed.

Houghton Takes Over

Several years before his selection as Secretary, Thomas B. Wales had moved his whole herd from South Framingham, MA, to Iowa City, IA, establishing it as the Brook Bank herd. Wales did much to make Holsteins known in the Midwest through sales and promotions. He set up the Holstein-Friesian Association office in his home in Iowa City where it remained until he dispersed his herd and moved back to Boston about 1891.

Mr. Wales, who had led the Association since its 1885 start, had antagonized

F. L. Houghton won control of the Association after a long battle with Secretary Wales.

many members by his actions; he ruled the Association with an autocratic hand, wielding his power through the use of proxies. Most members, unable to travel to the annual meeting, mailed a completed proxy back to the Secretary, who thus gained most of the votes, which then could be used at his discretion. At several meetings, Secretary Wales, using his overwhelming number of proxies, controlled every vote on every issue, causing widespread resentment among Association members. The time was ripe for a change in leadership.

F. L. Houghton, whose father Charles had been Secretary of the Chenery-led Association in 1871, used the power of the press to gain control of the reins of the Holstein-Friesian Association. In 1888, Houghton purchased the *Holstein-Friesian Register*, which had been started two years before in Indiana, and began using its columns to attack Wales. His writing persistence finally paid off when, at the March, 1894, annual meeting, in a bitter struggle, he defeated Wales on the eighth ballot.

Houghton set up his office in Boston but moved it in 1895 to his farm in Putney, VT, setting up shop in his farmhouse kitchen. The registry work was carried on from the Putney farm until 1903, when due to the business increase, he hired three clerks and rented office space in Brattleboro. Houghton, in a move that earned him the gratitude of hundreds of file clerks in the future, decided, in 1895, to change from handwritten record keeping to using a new machine, a typewriter, for the permanent animal record cards.

The Yearbook of the Holstein breed was started by Mr. Houghton in 1901 and published at his expense, by permission of the Board of Directors, until its publication was taken over by the Association in 1913.

The increase in business called for more office space and in 1917, A. B. Clapp, a Holstein breeder, built a handsome brick building on the southern edge of

Brattleboro, the same building which is now a part of the international headquarters of the Holstein Association. Patterned after a popular Dutch architectural style, the building featured parapets (now hidden by renovations) and graceful windows outlined in stone, from which one could see the Connecticut River. From these same windows, sharp-eyed employees could also spot the railroad train that carried Mr. Houghton to work each day from his Putney home. The Association leased the building from 1919 until finally purchasing it as a home office in 1928 for $30,000. The building is still an integral part of the Holstein Association operations.

Organizational Changes

When the Holstein-Friesian Association was first chartered in 1885, control was vested in a Board of Officers consisting of the president, four vice presidents, six directors, the secretary, treasurer and superintendent of Advanced Registry.

For nearly 30 years, the Association found itself constrained, under its charter, to hold annual meetings in New York State. In 1913, a committee consisting of former President O. U. Kellogg, chairman, D. D. Aitken, Fred F. Field, and A. L. Brockway, studied the situation and in the following year, the Association was re-incorporated under the Membership Corporations Law of New York State. The new charter allowed annual meetings to be held elsewhere. Consequently, in 1914, the annual meeting was held in Chicago.

The same committee on re-incorporation had also been asked to redraft the bylaws for consideration at the 1914 Chicago meeting. As a result of the committee's recommendations, an important change was made at that Chicago meeting, vesting the management and control of the Association in its nine-member Board of Directors. And in a move that surely contributed to the fiscal stability of the organization, board members were also made personally liable for debts and obligations created during their term in office.

The next organizational change of note came in 1919 when the Board of Directors was increased to 16, reflecting the growing size (membership had doubled in five years) and scope of the Association. There was a growing demand from different sections of the country for direct participation in management. The Board has remained at 16 members since 1919.

Production Testing Begins

One of the primary concepts that Solomon Hoxie, former Secretary of the Dutch Friesian group, brought with him to the Holstein-Friesian Association was that of an Advanced Registry (AR). This production testing idea, with inspection and production requirements, was started by Hoxie in the Dutch Friesian Herd Book. Hoxie and his group insisted that these procedures be continued in any unified Association, and they were. But the first tests for production were private, unverified by outside observers, and, therefore, subject

Malcolm H. Gardner was a proponent of year-round production testing.

to some skepticism. The tests were of short duration, usually 10 days, occasionally for 30 days.

It wasn't until the Babcock Test for butterfat was developed by Dr. S. M. Babcock of the University of Wisconsin, that the Association's production testing program began to grow. On March 15, 1892, Hoxie proposed that a new seven-day butter test be established, using the Babcock Test. But there was little interest (only 24 animals were entered in Advanced Registry in 1893) until, in a stroke of genius, W. G. Powell suggested to the 1894 annual meeting that the Association award prizes for authenticated weekly butterfat records conducted under the supervision of Advanced Registry. A sum of $1000 was established for prizes, and suddenly the AR program was off the ground. The AR records system was set up with supervision by state agricultural colleges and experiment stations, laying the foundation for an official testing system that has evolved into the DHIA/DHIR programs of the 1980's.

There was growing pressure to accept annual milk production records. When Solomon Hoxie resigned as AR Superintendent in 1905 at age 75, Malcolm H. Gardner, a proponent of yearly tests, was elected to replace him. Gardner, a Wisconsin breeder for 20 years, moved the Advanced Registry office from Yorkville, NY, to Delavan, WI. One of his first actions was to accept annual records. He immediately recognized and published records made earlier by college herds in Michigan in 1897 and Nebraska in 1902 and 1903. He also encouraged the University of Wisconsin to supervise semi-official yearly records on a two-day test each month.

Gardner was one of the most unusual leaders the Association ever has had —he was completely deaf from a childhood illness and was accompanied constantly by his wife, who relayed conversations to him using sign language. According to some who negotiated with him, one could never be quite sure that Gardner was getting the whole story. Mrs. Gardner, through judicious selection of dialogue, insulated him from many controversies, but, judging from the record, did not diminish his effectiveness.

Gardner's initiatives showed immediate results. By 1916, there were 12,882 cows on test, both official and semi-official (annual) testing. In 1917, the 305-day

Segis Pietertje Prospect was an early production leader.

test was inaugurated and grew in popularity. Cash prizes for AR records, having served their promotional purpose, were dropped in 1922, and soon participation in the seven-day test waned.

The Holstein Association had pioneered a whole new era of production testing as the first dairy breed to recognize and utilize the Babcock Test. The idea of Advanced Registry, so steadily pushed by Solomon Hoxie for years against industry-wide opposition, had finally been accepted by breeders. The spectacular advances of the breed in later years are in no small part due to production testing and the continued quest for higher-producing Holsteins.

Building A Field Force — The Beginning

Early Holstein promotional efforts came primarily from individual breeders and the Holstein-Friesian Association's Literary Committee, which had been established in 1895. The Committee began modestly, publishing technical articles and Advanced Registry test results in the agricultural press. At the turn of the century, articles by Malcolm Gardner titled "The Large Versus the Small Dairy Cow" and "Holstein-Friesian Milk for Infants and Invalids" enjoyed large circulation. The publishing efforts grew, and, in 1910, an advertising agency was hired for the first time. But after World War I, the Literary Committee, with a small staff and limited budget, was hard pressed to keep up with the faster pace and many breeders knew it. A change was needed to provide a more aggressive promotion effort.

D. D. Aitken established the Association's Extension program.

The mandate for action came from the 1919 Philadelphia annual meeting when delegates voted to raise the transfer fee from 25 cents to $1 to pay for a nationwide extension program. Association President D. D. Aitken, who was to become remembered as the "father of extension work," was hired to take charge of the extension effort working practically full time at an annual salary of $12,000 per year.

At the time, there were several state Holstein organizations, New York, Michigan, and Wisconsin, which were basically social groups with volunteer leadership. Aitken hired R. C. Pollock, former county agent from North Dakota, whose job was to encourage growth of state organizations. Membership campaigns were planned in all the leading Holstein states, and, as a result, New York, Michigan, Wisconsin, and Ohio were organized and soon were employing paid field secretaries. Next, Minnesota, Iowa, and Illinois followed, and a grass-roots field force was starting to build. The original state fieldmen organized by Mr. Pollock were: Chesley Jenness, Iowa; Elmer Zimmer, New York; Howard Barker, Ohio; R. M. Thompson, Illinois; Bertram Scott, Minnesota; Horace Norton, Michigan; and Les Oldham, Wisconsin.

Over the next 50 years, state fieldmen would, with the financial and technical support of the Association, play a major role in the growth and improvement of the Holstein breed.

After President Aitken got the program rolling, he turned it over to Fred A. Koenig, who served as Extension Director from 1921 to 1925. Koenig was

"Milk bars" were a major effort of the Holstein Association extension effort. Here the governors of Oregon and Washington toast their health at the Pacific International Livestock Exposition.

succeeded by Earl J. Cooper, who directed the program until 1931.

The fledgling state groups received a financial boost from the 1922 Kansas City convention delegates who raised the transfer fee to $1.50, with 50 cents of each transfer to be apportioned back to the state of origin. These transfer funds, which were to be the permanent funding source for many state Holstein groups for decades, were to be used under the direction of the national extension department. While the leading Holstein states had sufficient revenue to support their own field secretaries, others let the extension service apply the funds toward a fieldman for their area. By 1930, 21 states were using their transfer fees to support their own staff while 15 states were using fieldmen provided by the Association. A Holstein field force was in place.

Original Association Fieldmen
From the left: Chesley Jenness, Iowa; Elmer R. Zimmer, New York; Howard C. Barker, Ohio; R. C. Pollock, Organizational Director (under D. D. Aitken); R. M. Thompson, Illinois; Bertram D. Scott, Minnesota; Horace Norton, Jr., Michigan. Les Oldham of Wisconsin was absent. (The poster in the foreground was in wide use at the time.)

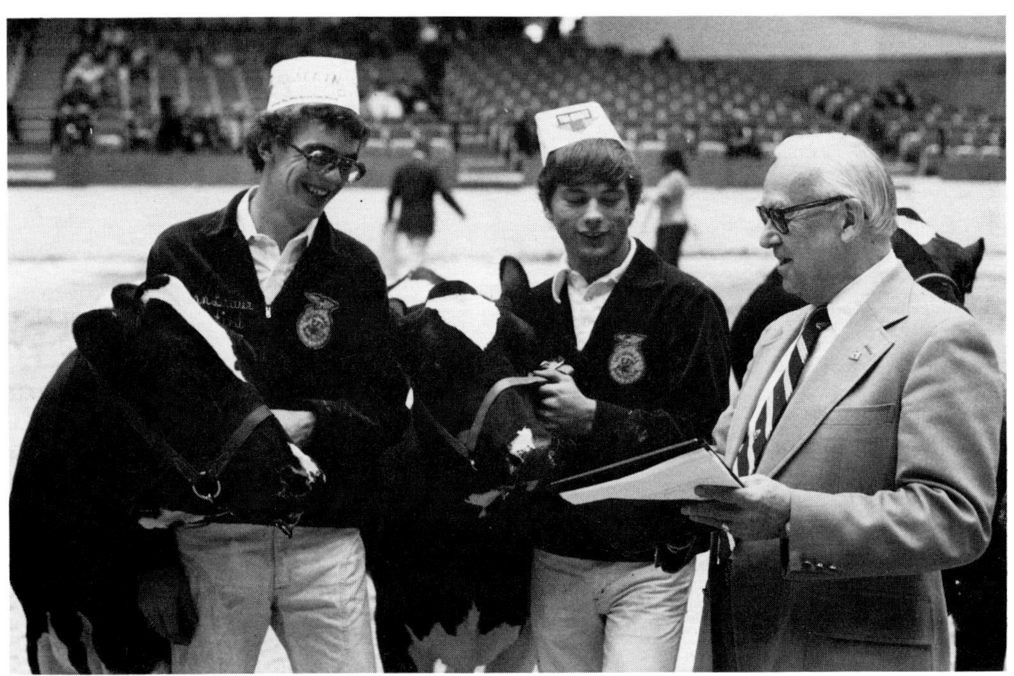

Association youth programs have continued since 1922.

Holstein Youth Programs

From the start of extension work, one of its most important components has been the programs for youth. In 1922, the Holstein-Friesian Association, recognizing that the future of the industry lay with youth on farms across the country, moved to actively support the 4-H Calf Club program. Earl J. Cooper was hired to the new position of Manager of the Junior Department and worked for three years building the program. Young people were encouraged to enroll in the Association, and, by 1924, 5225 youth from 28 states were enrolled. The Association, realizing that the program was, from the start, a grass-roots effort, encouraged competitive showing at county fairs by making calf club ribbons available to participants.

The most widely known Holstein activity for junior members is the Distinguished Junior Member Program which also was begun in 1922. Designed to focus national attention on Junior Holstein Association members, the program called for each state association to nominate two young people for the contest. Each year, a young man and a young woman were named National Champions, and eight finalists were recognized.

With over 10,000 junior members in its ranks, the Holstein Association modified the rules slightly in 1974 and now six equal winners can be named each year for this coveted award.

The Holstein youth program, supported by the Holstein Association, has remained an important function of state and local Holstein clubs, helping to promote the breed while drawing future industry leaders into the ranks of registered Holstein owners.

Past winners of the Distinguished Junior Member Program.

Maintaining Integrity — The Cabana Case

The first test of the Holstein-Friesian Association's authority came in 1885 (its very first year) when two Hollanders, the Sluiter brothers, applied to register cattle that had been imported and that were in quarantine. The Sluiters, after being denied, took the Association to court and lost. This action established the right of the Association to make restrictions on membership and on the acceptance of cattle for registration.

A long bitter battle was fought in the 1920's by the Holstein-Friesian Association to establish its right to govern its own affairs — specifically, to be able to cancel registration certificates issued as the result of fraudulent practices. The case centered around Oliver Cabana, whose herd at Pine Grove Farms outside Buffalo, NY, had achieved some phenomenal seven-day records. The herd was dispersed in 1919 at extraordinarily high prices, but as it turned out, many prices were fictitious and many animals never left the farm. Not only was the sale questionable, but the production records which stimulated the high prices were fraudulent.

Cabana's "herdsman extraordinary," Charlie Cole, who had made the phenomenal seven-day tests, moved on to eastern New York and Vermont after the sale and immediately began making a new series of high records, until he was caught. In July 1919, a test supervisor from the University of Vermont caught Cole red-handed adding cream and water during the milking from a rubber hot water bottle and hose worn under his clothes. Cole was brought before the Executive Committee of the Association and confessed, but later retracted the confession.

Cabana, who claimed that he had sold nearly $1,750,000 worth of cattle whose value depended on the authenticity of the records made by Cole, commenced suit in New York Supreme Court in Buffalo, enjoining the Association from cancelling the Advanced Registry Certificates. He argued that, in spite of what the bylaws might say, once the certificates were issued to him, they were his vested property and could not be taken by the Association, even if they were based on fraudulent records, as alleged. It was a challenge directed to the very foundations of the Association's authority, a challenge by a very influential individual. The

O. U. Kellogg, attorney for the Association in the Cabana case.

fight would continue for nearly a decade.

Judge Sears, in a ruling in early 1920, upheld the right of the Association to cancel records for fraud. Cabana appealed the ruling but the appellate division affirmed Judge Sears' ruling. Still not content with the decision, Cabana appealed to the highest court in New York, the Court of Appeals, which, on May 12, 1922, affirmed the appellate decision.

The records were cancelled at Syracuse, NY, on August 26, 1922 after a seven-hour hearing before the Board. Cabana, and several supporters, filed a succession of additional suits for the evident purpose of keeping the issue in litigation until the statute of limitations cancelled his liability. Thus, the issue was not completely settled until 1929.

The Association has been willing to invest in top legal counsel and that investment paid off. The Cabana case pointed out the value of having a strong attorney, in this situation, O. U. Kellogg, former President of the Association. C. M. Horn, who would serve admirably as Association counsel for many years, was the associate attorney for the Cabana case.

This landmark action in the Cabana case set precedents for not only the Holstein-Friesian Association, but other breed organizations as well. The most important aspect of this protracted litigation was the upholding of the right of an association of this nature to govern itself, to maintain the purity of its records, and make bylaws providing for the censure and expulsion of undesirable members. The long courageous battle by the Association maintained the integrity of records on file in the headquarters in Brattleboro; all owners of registered Holsteins over the next half century would benefit from the court victory.

The Proxy Fight

The Cabana case sparked one of the most bitter fights that the Association has endured. The struggle, centered on the use of proxies, resulted in a profound change in the operation of the Association.

When the Association was first established, each member had the right to vote, either in person or by proxy. This provision was to assure that there would be a quorum since the original bylaws called for votes by at least one-third of the membership. This quorum requirement was later dropped to 100 members but even that attendance was not reached until the 1904 meeting. So, while proxy voting may have been necessary to get business done, it also caused some acrimonious power struggles that jeopardized the vigor of the young organization.

As previously explained, in the early days, the Secretary sent out the ballots therefore receiving the majority of proxies. This practice, mastered by Mr. Wales, had led to his downfall in the 1894 election. Even though salaried officers could not hold proxies after 1893, Secretary Houghton also became a master at having enough proxies under his control (the proxies were simply mailed to his supporters rather than him) to overcome any challengers. This led to the first proxy battle, just before the Cabana case, in 1918.

The proposal to take the salaried officers out of politics (they were elected directly by vote of members) and make them appointive by the Board was, according to M. S. Prescott, who was at the meeting, aimed right at F. L. Houghton. A massive proxy campaign — "in defense of the faithful servants of the Association" — was launched by Mr. Houghton's supporters. It was the first meeting ever held in Wisconsin, and attendance was record-setting as 440 members participated. There were 6688 proxy votes submitted. The pro-Houghton forces controlled the proxies and the annual meeting.

The annual meeting of 1921 in Syracuse was the stormiest ever. The Cabana case was underway, and it had split the membership. Many who had purchased animals from Cabana feared serious financial loss if the records were cancelled, as President Aitken and some members of the Board wanted to do. (The three New York members, feeling especially vulnerable since the Association was chartered in New York, took a separate position opposite from the rest of the Board). Other members, even some of whom stood to lose money, felt that cancellation was the only proper course to pursue. The stage was set for the proxy battle of the century — in fact, the next-to-last proxy battle.

The campaign for proxies for the 1921 meeting resulted in four men holding 6890 proxies out of a total of 13,284 votes. They had campaigned on the promise of a reduced transfer fee, always a vote-getting issue, but it took three days to get the votes straightened out. According to H. W. Norton, who was on the proxy committee, "Many members had signed several proxies to different individuals, and we had to check all these and throw out all but those bearing the latest date."

The meeting, lasting two days, was full of oratorical fireworks regarding the Cabana case. But, amid the verbal pyrotechnics, in an action that was to change the archaic proxy system, Frank O. Lowden was chosen President. Lowden, who had been elected without being consulted, agreed to accept the presidency only if he had the full support of the whole Board, and, even more importantly, only if the proxy system be scrapped as soon as possible and replaced by a representative form of government.

A special meeting of the Association was called at once to consider changes in the bylaws to outlaw the proxy system. This precipitated a frantic fight for proxies which was won by the Lowden forces, and in St. Paul in October, 1921, the proxy system, after a stormy 36-year life, was finally buried.

True Type

As the proxy battles were ending, the Association was beginning a project called "True Type" that, in the words of H. W. Norton, "was one of the most important ever undertaken by a purebred livestock breeders organization."

After the show season of 1921, the *Holstein-Friesian World* encouraged a discussion about the problems in the show ring. A major problem seemed to be the lack of uniformity among judges' standards. Some judges liked one type of animal, while others had different preferences. In fact, some exhibitors carried a few extra animals on their show circuits in order to have the right animal for the right judge.

Axel Hansen of Minnesota, participating in the dialogue in the *World*, suggested that a conference of judges, breeders, and exhibitors be held, under the auspices of the Holstein-Friesian Association, to discuss a uniform idea of type to help standardize judging. Such a meeting was held at the Brentwood Sale, March 20, 1922, with Fred Pabst, Hostein-Friesian Association Director from Wisconsin, presiding. Secretary F. L. Houghton suggested that a set of official lantern slides of typical animals be prepared and distributed. Out of this idea

True Type Bull

Original True Type Mature Female

came the True Type oil paintings. Pabst suggested to the group that clay models be used to present the details of ideal conformation in a way that no score card or picture ever could. Thus was the True Type model concept born.

Immediately following the conference, the Association Board gave official status to the project by appointing a True Type Committee:

True Type Committee - 1922

W. S. Moscrip - Minnesota (Chairman)
R. E. Haeger - Illinois
Ward W. Stevens - New York
Prof. H. H. Kildee - Iowa
Axel Hansen - Minnesota
Prof. T. E. Elder - Massachusetts
W. H. Standish - Ohio
A. C. Oosterhuis - Wisconsin
Fred Pabst - Wisconsin

This committee of experts set out in 1922 to develop definite ideals of type in the form of pictures and models, examples that might be used by breeders and students as a standard when breeding for the future. The members scored and judged cattle at the Brentwood Sale and later, each selected photographs of animals that represented some particular feature of conformation that was the "ideal" for that part of a True Type animal. Edwin Megargee, painter, and Gozo Kawamura, sculptor, each recognized as a leader in livestock art, were hired by the committee. The job of designing an ideal animal was difficult, and only through painstaking work by each member did the task get completed.

By 1923, models of the True Type cow and bull, one-quarter life size, had been given to all the leading colleges in the country, and thousands of paintings had been distributed to breeders and agriculture schools. Miniature True Type animals were awarded as prizes at county fairs. The work of the committee had met with acclaim.

However, the True Type project had its critics as well. H. W. Norton, in his memoirs, tells of Holstein-Friesian Association Director Fred Field grumbling after the pictures of the ideal cow and bull were distributed: "a young man comes into my barn looking for a bull calf, and when I start to tell him of the merits of a certain young bull, he pulls a picture of the True Type bull out of his pocket and says 'that calf does not look much like this picture' and I have lost the sale of a bull." Even President Lowden was concerned about the groundswell of enthusiasm for type, reminding delegates at the 1923 convention that the quest for better conformation should not be accomplished at the sacrifice of production. It was a theme, a theme of parallel programs of breed improvement, that would be echoed by Association leaders for the next 60 years.

The True Type Committee did its work well. The animals they so carefully designed served the Holstein industry admirably, giving breeders a clear perception of the breeding goals for physical characteristics. Classifiers and judges across the world were trained using the models. The program would be expanded to include an "ideal type young female" in 1974 while the mature female True Type cow would be replaced in 1977. The bull model has remained unchanged. True Type, a revolutionary idea at first, eventually led to the classification programs of the Holstein-Friesian Association.

The Depression and War Years

The Great Depression tested the resiliency of the Holstein-Friesian Association. While the stock market crash of 1929 was not felt in agriculture for several years, by 1931, the price of registered cattle had plummeted. Association membership declined and financial resources were drained. Registrations dropped nearly 50 percent from 1929 to 1932 while transfers, always a good indicator of the health of the purebred industry, fell in 1933 to one-third the pre-1929 levels. The new breed improvement programs — Herd Testing and Herd Classification — begun so boldly in the late 1920's, floundered as the Depression deepened.

The era had begun in sorrow. Executive Secretary F. L. Houghton, who had helped build the Association from a small band of 422 members to a nationwide organization of 28,291, died in his office on December 19, 1927. He had just arrived at the Brattleboro, VT, office by train from his farm in Putney when stricken with a heart attack. His dynamic leadership over three decades is one of the highlights of the Holstein-Friesian Association's first century. Houghton was succeeded by his nephew, Houghton Seaverns, who had served many years as Assistant Secretary. Seaverns also would die in office, in 1938.

Houghton Seaverns

Horace W. Norton

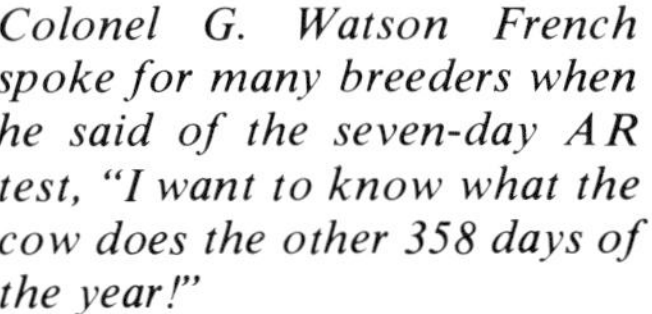

Colonel G. Watson French spoke for many breeders when he said of the seven-day AR test, "I want to know what the cow does the other 358 days of the year!"

The Advanced Registry department also had an important change in management when Malcolm H. Gardner, who had directed the department since taking over from Solomon Hoxie, retired in June, 1928 after 23 years of service. His successor, H. W. "Hod" Norton, would go on to lead the AR department through the Depression and the World War II years.

Complete Herd Testing (HIR)

The introduction of the Herd Improvement Test on January 1, 1928, was the beginning of the end for the short-term Advanced Registry tests on selected animals. The seven-day test, which under Solomon Hoxie had come into wide use in the early 1900's, was of little value in measuring the production capacity of a cow. However, the test had been an excellent breed promotion tool as high-producing Holsteins set new marks across the country. But it was time for change. Speaking out against advocates of the seven-day test, Colonel French of Iowana Farms, Davenport, IA, said, "I want to know what the cow does the other 358 days of the year."

Even the 365-day semi-official tests, which had started with several agricultural colleges in the early 1900's and been made part of Advanced Registry in 1908, recorded only the production of selected animals. For the same reason, the 305-day semi-official test, which began under AR in 1917, was an outdated tool. By 1925, there was growing interest among some breeders for a more realistic herd testing method. Not only were owners looking at the herd test programs operated by the Cow Testing Association under the Agricultural Extension Service (later known as Dairy Herd Improvement Association), they also realized that the Ayrshire Breeder's Association had adopted a herd test program in 1925.

Holstein herd testing was first considered at the 1926 convention at Des Moines when Peter P. Van Nuys of New Jersey offered a resolution calling for the adoption of a yearly test of the entire herd. The issue passed, and President Lowden immediately appointed a committee (J. Reynolds, H. W. Norton, M. S. Prescott, and AR Superintendent Gardner) to report to the next annual meeting. The 1927 report, while preliminary, was received with enthusiasm, and so an expanded committee, including professors with experience in the Cow Testing

Association, delivered a proposal that was accepted by the Board. The committee members were as follows: Reynolds, Norton, Prescott, Gardner, and Schaefer as well as professors Reed of Michigan, Wing of New York, Yapp of Illinois, Petersen of Minnesota, Borland of Pennsylvania, and Frandsen of Massachusetts. The testing system they designed would have a profound and lasting effect on the Holstein breed.

The herd test was a whole new concept from AR testing, in which owners had been able to select one or more cows to be tested. In the herd test, all females of producing age in the herd had to be tested — the herd became the unit and now, the producing ability of every cow in the herd was challenged. Herd testing required a different type of herd management from that where only a few top cows were tested. The program was received with skepticism by many, (when all cows were tested the herd averages appeared low to those used to the selective AR test results) and, in spite of encouragement from the Association, breeders were reluctant to switch. Dairymen were worried about the accuracy of the records for high-producing animals when mingled with a complete herd test. So a dual system, HIR and AR, went on for many years. HIR nearly did not make it; only 24 herds involving 4834 cows participated in 1929. Things went downhill from there as the Depression nearly killed the program, but it recovered in the late 1930's and would surge ahead in the post-World War II years.

The supervision and conduct of HIR testing, which was one of the strengths of the program with breeders, had a strong influence upon the development and acceptance of another herd testing program, DHIA, that was developing at the time. The seven-day AR test, which had fallen from usage in the 1920's, was officially dropped by the Association in 1932. However, the 10-month and yearly AR production testing would continue until 1965 but have declining participation as the herd test gained acceptance.

Herd Classification

Besides his dogged beliefs in Advanced Registry testing, one of the ideas that Solomon Hoxie brought with him from the Dutch Friesian Association in 1885 was the concept of classification — the physical inspection of animals before their acceptance into the herd book. Hoxie had designed a plan aimed at advancing the interests of cattle breeding, a system comprised of descriptions and measurements of herd book candidates. To his regret, animals were accepted into the Holstein-Friesian Association Registry based on ancestry alone — only imported animals were inspected. After the original imports ceased, there was no official classification program until 1929.

The herd classification program was a logical outgrowth of the True Type Committee. While the show ring brought some fine cows forward for inspection and comparison, relatively few breeders and few animals could be involved. An on-the-farm type evaluation and improvement program was needed. Now that breeders had an ideal against which to measure their cows, it seemed time to initiate a program to measure and recognize type against the ideal. But it wasn't easy — five years of study would pass before a program would be adopted.

The close tie between the show ring and classification is no suprise: all of the men who designed the program were show judges. And the same person who had led the True Type Committee, W. S. Moscrip, also chaired the group that designed herd classification. Helping him were:

Ward W. Stevens - New York
C. I. Miller - Ohio
R. J. Schaefer - Wisconsin
H. H. Kildee - Iowa
F. W. Atkeson - Idaho
J. B. Fitch - Minnesota
Frank L. Morris - California
Thomas E. Elder - Massachusetts

Moscrip, who would go on to be Association President, had been a respected cattle judge for years as well as the owner of the well-known North Star herd. Mr. Moscrip had only one good eye, his left one, which the earlier Holstein history reported: "...has probably had more to do with the type of the Holstein cow than any other single influence."

W. S. Moscrip, chairman of the "True Type" committee.

The first team of Association classifiers scores a herd in Illinois, February, 1929.

Pauline Beauty Johanna De Kol, first animal classified Excellent.

No sooner was the program underway when Moscrip got a chance to test his eye against other inspectors. In early 1929, the inspectors went to Illinois on the first classification trip. The men classified cows together, calling out the reasons for their choices, and defending their decisions. The inspectors found that they were in reasonable agreement over which cows were Excellent and which were Poor, but had great difficulty agreeing upon the demarcation line between other categories. This standardized approach to classification, begun on the very first farm visits, has been a key element in the Holstein Association's classification program since then.

The year 1929 was not the best time to start a new program, but even so, a total of 1957 animals were classified in 66 herds. Then the Depression hit and dairymen looked for ways to cut costs rather than to start new programs. Like herd testing, herd classification nearly perished — participation dropped to where, in 1934, only two herds were classified. The herd classification program wallowed until the early 1940's, when the pressures of World War II gave it new momentum.

Sire Recognition

It is no surprise that a sire recognition program began concurrently with the herd classification program. Unlike today, classification was initially a recognition program, a show ring for those who could not or would not transport animals.

The initial sire recognition program had Gold Medal, Silver Medal, and Bronze Medal Sires. The Bronze and Silver Medals were based on the type of daughters, and if the daughters of Silver Medal Sires met production standards,

North Star Gelschecola Champion, the breed's first Gold Medal Sire.

the sires then qualified for the Gold Medal Preferred Sire award.

The first sire to be recognized as a Silver Medal Sire was North Star Gelschecola Champion, who won the award just as the classification program began in March 1929. The inspectors, on the first classification trip through Illinois, judged his 25 daughters to include 10 Very Good, 14 Good, and one Fair. Since the daughters also had production records that exceeded the requirements for their age and class by at least 50 percent, North Star Gelschecola Champion qualified as a Gold Medal Proven Sire.

The system was changed in 1945. The Bronze Medal rating was discontinued and a Silver Medal Type Sire award was added. The award was no longer based on the recommendation of the inspector, but on the average score of daughters. At the same time, a new rating, called the Silver Medal Production Sire was adopted. If a bull got both Silver Medal Type and Silver Medal Production, he then qualified for the Gold Medal Proven Sire.

The Gold Medal Sire program award has periodically been changed to take into account changes in the industry. The 1985 requirements for Gold Medal Sire are based on the genetic performance of the bull. The award is now a permanent recognition.

Progressive Breeders Registry

Another recognition program, Progressive Breeders Registry, began in the 1930's. H. A. Mathiesen, Holstein-Friesian Association fieldman for the western states in the early 1930's, came up with the idea to honor dairymen who not only were leaders in promoting the Holstein breed, but also had high producing herds. His successor as western fieldman, M. B. Nichols, worked to spread the idea and in early 1937, four state Holstein clubs (Idaho, Washington, Oregon, and Colorado) adopted the program. Soon it caught the attention of Secretary Norton who asked Nichols to describe it to the 1937 Association convention. The PBR system was adopted in 1938 and the first awards, to nine breeders, were made in 1939.

PBR was designed to stimulate interest in the Association's lagging breed improvement programs by recognizing breeders who used production testing and

In 1939, the Association consolidated all operations in its Brattleboro, Vermont office.

classification and kept their herds disease-free. The requirements for the program have remained stringent and over the years, the PBR award has been a valued accomplishment for the breeders and members who have met its high goals. Since the program began, 1283 plaques have been awarded to first-time winners while many others have qualified for successive years. As of 1984, the University of New Hampshire had received the Progressive Breeders Registry award for 40 straight years.

A Permanent Home For The Association

The Association purchased the Brattleboro, VT, office in 1928, having rented it for a decade. It also built a modern building in Madison, WI, in 1931 to house the Advanced Registry and extension departments. But the financial pressures of the Depression called for action — both buildings could no longer be afforded. The situation raised again the issue of where the headquarters office should be located. Breeders in the Midwest, especially Wisconsin, pushed to have the Association move to the heartland of the dairy industry. Unfortunately for them, the Madison property was too small to house the complete operation but being newer and smaller, was easier to sell. The Board consolidated all operations in its Brattleboro office in April, 1939, and sold the Madison property at a favorable price. And they were welcomed by the small Vermont town. At a special town meeting on May 27, 1937, the Association was exempted from real estate and personal property taxes for 10 years, beginning April 1, 1940. The Holstein-Friesian Association had established its permanent home in the Green Mountain State.

The Association Weathers The Storm

The Depression put severe pressure on the Holstein-Friesian Association. Not only did it put the organization to the financial test, it nearly killed the two fledgling breed improvement programs: herd classification and herd testing. Both would be saved by the outbreak of World War II and the resulting demand for Holsteins.

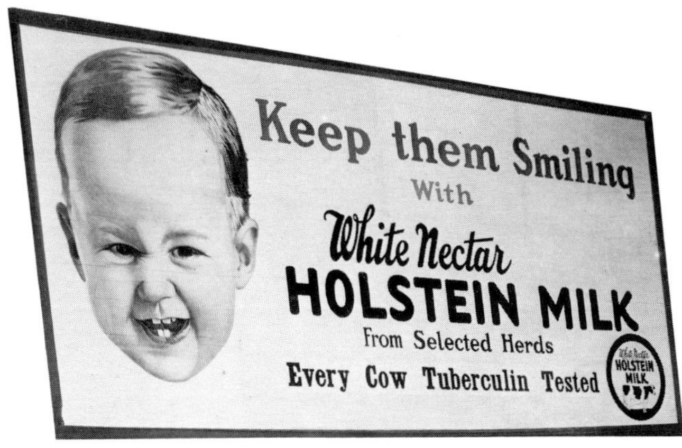

The "White Nectar" milk marketing campaign of the 1920's was a failure.

Another program, a well-conceived milk marketing effort, never made it off the ground. This was the second try at promotion: back in the 1920's, the Association had launched an abortive attempt at marketing Holstein milk under the trademark of "White Nectar." Unfortunately, production sources were not in place and in spite of a massive expenditure of $103,892 in 1923, the program flopped. But "Holstein Holsum Milk" was to be different — a marketing program designed from the start with the producer in mind. In one of his first projects as Director of Extension, Glen Householder had planned the "Holstein Holsum" project in the late 1930's to attack the marketing problems of the Depression. He got the program approved in 1940 and was lining up producers (61 had signed by the end of 1940), when the entry of the United States into the war made a milk-marketing program superfluous and it was discontinued.

While these programs felt the crunch of the Depression, the very financial base of the Association was threatened as year after year of operating losses occurred. Fortunately, early leaders had realized that fiscal stability was the key to shoring up the lineage records of the Association. The Reserve Fund, established in 1902 as a "backlog for years of less activity," helped keep the Association solvent during these trying times.

Back in 1916, at a time when the reserve balance stood at about 100 percent of the year's expenses (there was a one-to-one coverage), Association President Aitken discussed the Reserve Fund in his report to the membership. Aitken wrote: "The amount is no larger than correct business policy demands that we should have, as a safeguard against a possible long period of depression." His words were right on target — the depressed economy of the 1920's and 1930's forced the Association to use these set-aside funds to stay solvent. Draws in the 1920's were small, but during the Great Depression, seven straight years of losses (1930-1936) drained $155,000 from the Reserve Fund, leaving it at $101,000 or only 50 percent of annual operating expense. But, the foresight and the frugality of the past had paid off. The Holstein-Friesian Association had weathered the financial storm and survived the 1930's.

Holstein breeders were part of the "Food For Freedom" program launched in the early days of World War II.

Wartime Growth

World War II changed the course of the Holstein breed and the Holstein Association, triggering an expansion that would continue for the next four decades. Heavy demand for milk products brought the high-producing Holstein cow to the forefront, and for the first time since 1929, the Holstein breeder was in a developing market with just the animal needed to meet the demand. Concentrated feed was scarce and the Holstein cow showed her inherited ability to convert farm-grown feed to milk. There was a severe labor shortage and dairymen soon learned that 20 heavy-milking Holsteins took no more help than 20 common cows. Those who had struggled through the Depression years to maintain herd-building programs were rewarded: their black and white animals were in demand.

With the entry of the United States into the war, the nation launched a "Food For Freedom" program. Association Secretary H. W. Norton, Jr., writing in 1942 to the members, issued the call: "....Our sons will be out there on the battle line fighting for us, and we who remain at home must do our utmost at whatever task we may be issued. We must be part of the great ring of supply, and support our men in the line of battle...." He wrote of the opportunity that Holstein-Friesian breeders would have to serve their country when dairy supplies were so essential to victory, noting that the average production of dairy cows, according to USDA, was 4575 pounds of milk, while the average Holstein on the Herd Improvement Registry Test produced 11,859 pounds.

For Holstein breeders, the major development of the war was the decision by the Army Food Procurement Division, after recommendations by Public Health Service nutritional experts, to base the buying standard for milk products on 3.5% milk. Suddenly, years of "creamline" arguments were null and void. Glen M. Householder, Director of Extension Service, gave this explanation in the

Minnow Creek Eden Repeat was a production leader of the 1940's.

Association's 1941 Annual Report: "With shipping space at a premium, and in selecting foods that would make the greatest contribution to Britain's nutritional defense, those ingredients in milk which lie below the cream line were selected."

The contention of Holstein breeders that there was more to milk than butterfat was confirmed. It was a decision that changed the fortunes of Holstein owners ever since, and led, in later decades, to a recognition by consumers all over the nation that the minerals and vitamins in milk were vastly more important to human health than butterfat.

The needs of the war effort drove scientists to accomplish technical breakthroughs in the dairy industry as in the defense industry, achieving results in months instead of years. For example, the constraint of "perishability," which had limited the spread of milk products into the areas of the country and the world that lacked refrigeration,was overcome by advancements made in the development of dried whole milk. GIs in trenches and foxholes all over the globe were drinking reconstituted milk, much of it Holstein-generated.

The war also had an effect on the market distribution pattern. As thousands of training camps sprang up around the country, particularly in the southern states, distribution patterns for milk changed. This new market for milk prompted southern farmers to eye the many abandoned tobacco and cotton farms with the thought of starting dairy farms. The Holstein Association recognized the potential market for registered and high grade Holsteins and hired a new fieldman, Thomas E. Elder, who brought years of experience as a judge, educator, and breeder to the Association's staff, to open the southeast market for Holsteins.

Meanwhile, on battlefields and military posts across the world, thousands of American servicemen dreamed of returning home to start a farm. They wrote to the Holstein Association for guidance, and many, when they got back, brought new energy into the Holstein field programs and the fledgling state association programs (Chapter 3).

The Field Force Expands

Since the start of the allotment program in the 1920's, the smaller Holstein

Holstein fieldmen (now called consultants) have been helping breeders for over 40 years.

states, which received relatively small allotments, had found it difficult to support full-time state fieldmen. Faced with this financial reality, in 1935 five more states (Michigan, Wisconsin, Minnesota, Kansas and Missouri) were added to the area covered by the national extension work, which by now covered nearly half the Holstein-Friesian Association membership. By 1940, only New York, Ohio, Pennsylvania, Iowa and Indiana were maintaining state fieldmen.

The national fieldmen had vast territories to cover: Allen Crissey covered the Eastern Seaboard from Maine to Virginia; E. M. Clark drove the roads of Missouri, Illinois, southern Wisconsin, and lower Michigan; Robert Geiger handled upper Michigan, northern Wisconsin, Minnesota, South Dakota, Kansas, and Oklahoma; M. B. Nichols had 11 western states; and R. L. Pou served eight southern states. Another service area was created in 1941 when North Dakota and Nebraska dropped the allotment system. Elmer A. Dawdy of Salina, KS, was hired to serve as national fieldman for a massive new territory consisting of Kansas, Oklahoma, Texas, New Mexico, Colorado, southern Wyoming and Nebraska. By the late 1940's, only New York, Ohio, and Pennsylvania were still operating independent field operations, a situation which would continue until the 1970's.

National fieldmen, spread thinly across the country, spent long hours driving to meetings and shows to spread the word about Holsteins. But as World War II went on, they found that their role was changing as the popularity of Holsteins caused breeders to realize that new methods were needed to buy and sell cattle. Restrictions on gasoline and tires were making the buying and selling of Holsteins by mail more commonplace. No longer could prospective buyers drive miles to personally check animals before purchase. Often it was the word, or the score, of the Association classifier that helped make the sale. The herd classification program took on new meaning (9002 animals were classified in 1942, three times the 1941 count) as breeders began to recognize the advertising value of classification scores.

Holstein-Friesian fieldmen, also hindered by travel restrictions, became scouts for breeders, helping to locate the animals needed to increase production. This was the advent of the Association's field program, foreshadowing the roles that

World War II strained the operations of the Holstein Association office.

future fieldmen would play — roles that would provide hard, practical, on-the-farm selection, culling, and breeding advice.

Problems At The Office

Millions of palates had grown accustomed to moderate-fat milk during the war. Consumers liked homogenized milk in the new cardboard containers and milk plants needed fluid milk more than ever. It was a situation made for the Holstein dairymen. For years, the breeding program of the Holstein-Friesian Association and Holstein breeders had been aimed at high milk production and moderate butterfat. Holstein dairymen, whose predecessors had suffered during the first half of the century when the emphasis was on high fat level, were now in a position to take advantage of the changed situation.

But the postwar period also had brought competition back to the milk marketplace. Gone were the wartime subsidies and "produce at any cost" procurement programs. The time was ripe for an efficient, high-producing cow. The Holstein-Friesian breed had expanded sharply during the war and now, 44 percent of all cows registered were Holsteins. And the rush was on to register more.

The new popularity of Holsteins caused problems. As the war ended, hundreds of dairymen realized a new value in their Holsteins and rushed to reinstate herds whose registrations had been allowed to lapse. Consequently, an avalanche of paperwork buried the Brattleboro office of the Association. Postwar 1946 was the most profitable year for the organization to that date, but it took its toll. The Association, in a Herculean effort, brought the immense backlog of registration and transfer applications nearly up to date, but the catch-up effort, while generating large revenues, was accomplished at the expense of many of the other services of the Association. It was a classic case of "crisis management."

Internal operating problems had been developing. Continued delays in processing the workload and getting the herd book published resulted in backlogs that seemed to be getting worse. All the old applications for registry (since 1885) had been kept on file and were clogging the limited storage space. The war industry had drawn people from their normal occupations, sapping the talent of the Association. By 1943, 85 percent of the office work force had less than a year's experience. This loss of personnel, coupled with ever-increasing registration and transfer requests, strained the operations of the Brattleboro office. The Association ended the war the way it started, with a serious backlog problem. By March 29th, 1946, there were 70,000 applications waiting for attention.

Secretary Norton, writing with his typical candor, said, "Our members throughout the country are annoyed and concerned with delays in the business through the office." And the situation continued to worsen, forcing sweeping procedural and management changes in the Holstein-Friesian Association.

The Efficiency Study

The Brattleboro office facilities of the Association were taxed to the limit in 1946. Every available square foot in the 1919 building was converted to office use. Added office facilities were being rented on the outside along with stockroom and storage space. The organization was simply outgrowing the building — there was an immediate need for space and greater efficiency.

Paperwork swamped the office: requests went unanswered for months, employees were disgruntled (80 employees, half the work force, left during 1946), and the field staff was hampered by the delays. The Association was missing "the forest for the trees," being continually inundated by the stream of paperwork. Management decided, because of the shortage of help, to stop preparing pedigrees and consequently, the service was picked up by pedigree houses. Production records and herd classification data lay buried and unprocessed in Association files, unavailable for breeders and others to use.

Secretary Norton said of the problem: "Of late years our organization has not kept pace with the breed. Our vision has been too obscured by immediate obstacles to look into the future and plan for things to come...."

By 1947, the operating problems were reflected in the financial statement. The Association was forced for the first time since the Depression to dip into the Reserve Fund to cover a deficit.

Action was taken to head off disaster. In a move that had far-flung effects, the delegates to the 1947 convention amended the Association bylaws to limit the terms of the Directors of the Association to two consecutive four-year terms. Gone were the long-term memberships on the Board. An infusion of "new blood" on the Board seemed to be the first step to answering the problems.

Taking a cue from the delegates, the Board of Directors responded to the operating problems by authorizing the employment of a management firm to make a thorough study of the Association. A committee of Carl Wooster, President, Harold J. Shaw, Chairman of the Executive Committee, and Albert Craig, Chairman of the Finance Committee, oversaw the management study. They hired Barrington Associates of New York who studied the management structure, the facilities, and the operating procedures of the Association, and then made a complete report and set of recommendations to the Board.

Since its early days, multiple leaders, sometimes as many as four in offices in different sections of the country, ran the Association. Even after all offices were combined in Brattleboro, the practice of multiple leadership persisted. Each section was autonomous, and there often was a lack of coordination, if not outright competition, each with the other. As volume had grown, management by committees was not working. In their most important suggestion, one that would change the Association for decades to come, Barrington Associates advised the Board that the whole management system should be scrapped and changed to an executive management system, a change as revolutionary as the proxy vote change three decades before.

The consultants acknowledged that the relationship between members, the Board, and management "must be on a different plane than in a commercial institution" and agreed that the membership and the Board would, and should, take an active role. However, they recommended, for efficiency's sake, that all operations of the Association be directed by an Executive Secretary, who would rely on committees of the Board for counsel and advice. One person would be in charge. This major change, from committee management to executive management, would be the first step toward a more business-like approach of operating the nonprofit corporation.

The Association Reorganizes

The consultants submitted their report to the December 1947 Board Meeting. It was accepted and immediate steps were taken to implement the proposals. The delegates at the 1948 convention in Kansas City approved the new management plan through extensive revisions to the bylaws.

The reorganization was completed in 1948. The Board dropped any operation responsibililty and took on a policy-making role. Under the Executive Secretary, Mr. Norton, there were five principal divisions:

Extension Service Division	Glen M. Householder
Registry Department	Keith Watkin
Office Management	A. H. Winchester
Advance Registry	Leo R. Blanding
Accounting	Sedley F. Dunlap

One of the first actions taken by the Board was to have a permanent effect on the future of the organization. On October 1, 1948, Robert H. Rumler, a young executive from DuPont with a degree in dairy husbandry and a prior background in county extension work, was hired as Assistant Executive Secretary. Many of the advances of the next 25 years would be the result of the strong leadership that Rumler brought during a long tenure as Executive Secretary.

As the 1940's ended, the Holstein-Friesian Association had the new management structure in place. In spite of continued crowded office conditions and no additional employees, work output and efficiency immediately increased under the new system. The Association had organized itself for an upward charge through the 1950's.

Getting Down to Business

The fabulous fifties. Every year seemed to set new records in registrations and transfers as Holstein Association programs grew and flourished under the new management system. The annual reports from the period all have the same theme: "The best year ever." It was a time that saw the end of the widespread use of the neighborhood bull as artificial insemination changed the dairy industry. At the same time, "Holstein fever" spread across the country as state associations continued to spring up and an expanded Holstein field staff got out and spread the word. During the 1950's, the ratio of registered Holsteins to the nation's dairy cows doubled: in 1950, the registration of Holsteins reached 184,246 or 46 percent of the dairy cows registered in the country. This meant one Holstein registration for every 119 cows in milk. Ten years later, registrations at the Brattleboro office would reach 265,861 — 64 Holsteins would be registered out of every 100 dairy cows registered. During the same period, there was a decrease of 4 million milk cows across the country, so for every 66 cows in milk, one Holstein was registered. It truly was a fabulous decade for the Holstein-Friesian Association.

Changes At The Home Office

With the management sails trimmed for action, the Holstein-Friesian Association launched into the fifties with two major projects to improve productivity: introduction of the first computer system and a major structural addition to the office.

After careful study and numerous surveys by the building committee, a special meeting of the Board in Des Moines on October 3, 1950, considered and approved

Workmen install the Association's first computer.

a 50′ x 90′ building addition. The construction contract for the building, which would nearly double the office space, was awarded to William Cushman & Son of Brattleboro. Tons of Vermont ledge were blasted to excavate the site and the addition was successfully completed, within the budget, in 1952. Although the Board had authorized the withdrawal of funds from the Reserve Fund, the total cost was paid from operating revenues of the Association.

The first computers were purchased, under the direction of Dr. George Barrett, who was Superintendent of Advanced Registry at the time, during the same period. IBM punch card equipment was installed in 1952, and in 1959 IBM 826 machines were installed in the Registry Department. As certificates were typed, IBM cards, used to produce the animal record cards as well as the herd book, were automatically created. Three separate typing and proofreading operations were combined into one. Mechanization and revised operational procedures meant fewer clerical workers: with the new system, a clerk could handle three times the number (135-145) of registration and transfer certificates in a week than under the old system. For the first time in 10 years, the publication of herd books was timely. The new executive management system was getting results.

H. W. "Hod" Norton, Jr., the man who steered the Holstein-Friesian Association through the war years, retired in December, 1953. Norton, who often had worn a number of Association leadership hats at the same time, had devoted most of his life to the Holstein breed. He had served as Secretary and later as Executive Secretary since 1939 and also had been the Superintendent of Advanced Registry for 12 years prior to that. He was another in a legacy of leaders who have led the Association to a position of prominence. During his tenure, the Association reorganized, expanded the office, and mechanized the operations. He became Chairman of the newly-formed Research Committee.

On December 4, 1953, Assistant Executive Secretary Robert H. Rumler officially took over the reins of the Association and led it on a 25-year trek that would take U.S. Holsteins to all corners of the country and the world.

H. W. "Hod" Norton

Robert H. Rumler

The Phenomena Of AI

The tremendous growth of artificial insemination (AI) during the 1950's had a major impact on the Holstein Association. As the new decade began, AI, which had started on a small scale, had reached the point where a quarter of the animals being registered had been conceived artificially. At the end of the period, this percentage was close to 50. The continuous growth put new pressures on the Association to keep the registry records credible. Fortunately, the parallel development of blood typing during the same period helped the Association meet that challenge.

Enos J. Perry introduced artificial insemination to American dairymen, having learned about it in Denmark during a sabbatical tour in 1937. When he returned later that year, Perry, who served as an extension dairy specialist at Rutgers University for 33 years, proposed the new technique to the dairy farmers in Hunterdon County, New Jersey. The first inseminations took place on four farms on May 16, 1938, and soon, over 1000 cows on 97 farms were enrolled. Initial financial support for the project came from the New Jersey Holstein Association. The first AI calf was born on the Stanton, New Jersey farm of R. S. Schomp on February 15, 1939. To the surprise of some, she was a normal animal in all respects. The days of leading "Old Maggie" down the road to be bred by the neighbor's bull were nearly over.

AI grew slowly during the war years, but still posed a real challenge to the Association. While there were differences of opinion as to its value, the technique was expanding fast in the commercial herds, and gains in production, due to the more extensive use of better sires, were starting to occur. Registered breeders were worried: first, that the production superiority that they had always enjoyed might be lost, and second, that they were losing one of their prime markets — the sale of young bulls.

At the beginning of AI use, Holstein bull registrations accounted for about 31 percent of all registrations. By 1950, the growth of AI had resulted in a downturn in bull registrations to 20 percent as the number of sires in service was reduced. By the end of the fifties, it had dropped to 10 percent of total registrations, continuing downward in the next 10 years to 7 percent. A further decline has

Enos J. Perry

taken place over the last 15 years to less than 5 percent. The breeders' fears in the early fifties about bull sales had been justified, but the genetic advantages that would be achieved through AI would make those fears a forgotten issue.

Many purebred breeders, traditionalists at heart, were slow to start using AI. In 1947, only 13 percent of the registration applications were from artificial insemination. But, by the mid-1950's, the process of freezing semen had become more common and its growing use forced any breeder who wanted to stay competitive into using AI. Now a single sire could have thousands of daughters; the dairyman's choice of top sires was limitless. By 1959, over half the calves registered at the Holstein office were the result of artificial insemination and nine calves a day were being registered from one popular sire.

One of the major concerns about AI was the fear that great harm would come to the breed by using some weak bulls. The potential impact of a single bull, mated to thousands of females, was being recognized. There was a real need to find and prove genetically superior sires, bulls with the ability to transmit the desired traits. But the evaluation tools were inadequate to meet the demands of

Wis Burke Ideal was one of the first sires used extensively for artificial insemination.

AI. The Association recognized, along with others, that procedures for proving bulls were not as dependable as desired and urged caution. Many questioned the wisdom of accepting proof based only on five dam/daughter comparisons but there was little else to use. The groundwork was being laid for major advances in production testing, type classification, and the processing and use of such data —advances that would occur in the sixties and result in dramatic genetic advances.

Keeping Records Credible With Blood Typing

As the registration and transfer business grew, as the value of individual animals increased, and as the genetic contribution of a limited number of sires intensified, so the need for increased accuracy of Association records became more critical. The cornerstone of credibility that had been built over the last 65 years had to be reinforced. Blood typing gave the Holstein Association the tool it needed to control the accuracy of the Holstein ancestry records.

In a coincidence that had a major impact on breed improvement and the credibility of records, the procedure of blood typing was developing during the same period as AI. Drs. Irwin, Stormont, Ferguson, and Rasmussen recognized, in the late 1930's, the possibility of blood typing cattle, using the principle of immunology. Dr. E. E. Heisner presented the process to the Holstein-Friesian Association in June 1939 which, along with the American Guernsey Club, made a grant to support and develop the program. During the war, blood typing grew slowly. In 1947, when Dr. Ferguson joined the staff at Ohio State University, an offer was made to the Purebred Dairy Cattle Association (PDCA) to conduct a blood-typing program and the modern-day program began in earnest, just in time to allay the fears arising over artificial insemination.

During the early years of AI, there was fear among breed organizations that it would be difficult to determine whether a particular bull's semen was or was not used to breed a particular cow. The rules for registry from AI were cumbersome and difficult to administer and imposed added responsibility on those using AI, but served to protect the accuracy of records. Yet there was a need to do more to protect the credibility of records—even the most conscientious breeder could err

Richard E. Nelson

in judgement as to which of several bulls could have sired a calf. Recognizing this, the PDCA, in which the Holstein Association was a leading force, developed uniform requirements in 1948: "All bulls in bull studs or used between herds must be blood-typed as soon as service is available."

During 1949, 523 Holstein bulls were blood-typed at the Ohio State lab bringing the total completed to over 1200. During the same year, the Board gave instructions to start spot checking calves for parentage, a pedigree accuracy check that has continued for 35 years.

Beginning in 1950, the Association required that all bulls used in the AI program be blood-typed in order for their progeny to be eligible for registration. Blood-typing service testing was shifted in the mid-1950's to the Serology Laboratory of the University of California at Davis.

The utilization of blood typing was a key ingredient to the success of the fifties as the Holstein Association, in an era of rapid growth and increased breeding sophistication, kept its genealogical records credible. By the end of the decade, when over half the applications for registration were for artificially conceived animals and thousands of samples of frozen semen were being used all over the country, the new blood-typing technology added immeasurably to the level of accuracy and credibility of the Association's herd book records. Blood typing had provided the needed guarantee.

The Holstein Association, a leader in the blood-typing field from the start, has continued a rigorous program under the conscientious direction of Richard Nelson. Nelson's personal dedication has made the program a model for breed organizations and resulted, in recent years, in the Association managing the blood-typing programs of other livestock breeding groups.

Parallel Breed Improvement

The two breed improvement programs, production testing and classification, grew steadily during the years after the war. Herd testing programs for production grew in popularity while Advanced Registry testing, which had reached a peak in 1950, steadily declined. Herd classification, which got off to a slow start during its first dozen years, continued the growth that had started in the

Herd testing programs for production grew in popularity in the 1950's.

late 1940's. Breeders were starting to recognize that type information combined with testing results were the key to successful Holstein breeding.

Herd testing (HIR), which had been received with skepticism by many breeders when it was first introduced in the late 1920's, continued to grow in popularity over the AR testing program. By the mid-fifties, nearly 75,000 cows were enrolled on HIR test. In 1956, the Association changed from the old herd average plan to the use of Mature Equivalent lactation averages, converting all records to a mature, 2X basis, putting each herd on an equal basis.

In the same year, in a major breed-improvement step, the Holstein-Friesian Association accepted DHIA records, on a herd basis, for the first time. DHIA was becoming computerized and more widely used by Holstein dairymen. Recognizing that the program would help members, the Association moved to have those records available for use and publication by the Association. The DHIA records were used in all Association breed-improvement programs including daughter/dam comparisons, lifetime production, Progressive Breeders Registry, and sire/dam recognition. The 305-day records were published in the Type and Production Yearbook. The handwriting was on the wall for the AR test, but it would doggedly hang on into the next decade. The Advanced Registry Honor List was dropped in 1958. There were now two herd-testing programs available for Holstein breeders, HIR and DHIA and it was obvious that the efforts would have to be coordinated. That would happen with the introduction of DHIR in the coming decade.

The herd classification program continued to snowball as more and more breeders recognized the value of an impartial appraisal of the type and conformation of their animals. Up until now, the classification program had been conducted when part-time classifiers were available and a "reasonable" number of breeders could be encouraged to classify at a time of their preference. In what

Carnation Homestead Daisy Madcap became the first 1500-lb. butterfat producer in 1957.

The employment of full-time classifiers by the Holstein Association was an important step in breed improvement.

turned out to be one of the most important breed-improvement moves of the fifties, the Association ended the system of part-time classifiers and put classification on a year-round, regularly scheduled basis. Two full-time employees began work in the mid-fifties and by the end of the decade, five experienced men were classifying herds across the country on a full-time basis. This landmark action in the classification program signaled the start of a force of professional, career field personnel in the program.

State And Local Holstein Clubs

Grass-roots activity has always been the key to Holstein promotional efforts. Executive Secretary Zane Akins explained it this way in a 1985 interview: "The momentum and activity strength of the Holstein Association is in the local organizations. We've supported them by giving state members a discount on registration fees. Over the years, a strong tie has been developed."

The effectiveness of the national extension program has always been linked with the efforts of the Holstein groups at the state and local level (At first, these

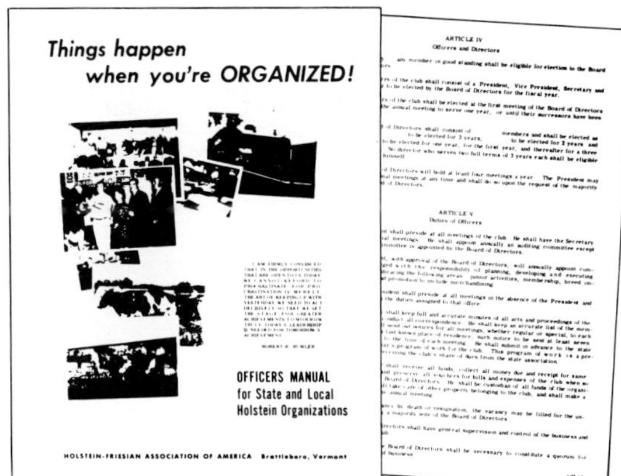

Many state and local Holstein groups received support and guidance from the Association during their formation.

groups were the Association's extension effort). Starting as a way for breeders to get together to promote dairy cattle, the clubs combined business with sociability, bringing together people with a common interest—the Holstein cow. As World War II ended, the groups were rejuvenated by an infusion of energy and leadership from many returning GIs. By the start of the fifties, nearly 400 Holstein clubs were carrying out educational and extension programs. And while the hiring of five full-time classifiers was a major step in spreading the effectiveness of the classification program, the steady move toward permanent Association fieldmen supported the continued growth of the extension effort.

When the Florida club was organized in 1954, there were clubs in all 48 states, as well as 426 county clubs and 11 junior Holstein clubs. The enthusiasm and promotion activity of these field groups contributed greatly to the remarkable growth in classification, production testing, and other programs of the Association during the 1950's and 1960's.

As the revitalized state and local Holstein organizations stepped up activity, the Holstein Junior Program shattered previous records. Each year in the late 1950's, more students were accepted as junior members — 1927 in 1957; 2086 in 1958. By the end of the decade, 15 Association fieldmen were working with over 500 Holstein clubs across the country. From the "Black and White Shows" —which were state and local Holstein group projects from their inception — to locally sponsored consignment sales, a network of Holstein promoters was in place and working. The Holstein-Friesian Association was helping to provide the direction and initiative needed to help the state and local clubs promote Holstein cattle.

The Field Force Builds

By the mid-fifties, there were 15 professional fieldmen working for the Association: only New York, Ohio, and Pennsylvania were accepting an allotment of transfer fees to help support their own field staff. But, with field staff in three large Holstein states working independently, the coordination of the extension effort was difficult and a continued concern to Holstein-Friesian Association leadership. An attempt to bring more cohesion to the extension

Fieldman Bob Cain spread the word about Holsteins at many fairs and shows in the early 1950's with this mobile exhibit.

effort took place in 1958 when the 50 cents per transfer allotment was discontinued. The three states entered into a closer contractual and operating arrangement with the Association, a system that worked better but was still unwieldy. Not until the mid-1970's would the effort be uniform, with an Association-directed field force serving all states.

This concern over coordination was more than just trying to build an Association field force — uniformity was crucial to the success of the national extension effort. A key element was the coordination of the activities of the state and local Holstein clubs. It was essential to get these grass-roots teams all working toward the same goals, and one of the jobs of the fieldman was to help direct these teams. Therefore, it was imperative that all fieldmen were giving similar directions in the field.

Holstein planning conferences, begun in 1945, helped get that essential job done. Held in each Association field territory, they brought together representatives of state Holstein clubs, the area fieldman, a college dairy specialist, and national office extension staff to review the program of work being carried out and to map out the development program for the area for the coming year. This method of obtaining feedback would eventually mature into Winter Forums in the 1970's.

More standardization was achieved when the annual fieldmen's conferences were begun in 1953. All 15 Holstein fieldmen gathered in the Brattleboro home office to discuss mutual problems and to highlight individual successes. These conferences became an annual planning and standardization affair, setting the tone and emphasis for the coming year. It gave the headquarters staff a chance to work with the fieldmen on "special emphasis projects" and contributed greatly to the coordination of Association programs in the field.

The growth of the state and local Holstein groups during the decade placed new demands on the time of the fieldman. Suddenly, everyone wanted him to attend night meetings. In fact, the last session of the planning conferences was spent carving up his time and travel, on a general schedule, among the participating

Fieldmen's conferences began in 1953. Here are the attendees at the 1957 session in Brattleboro.

states. This dependency upon the fieldman for meetings and program-presentation by local and state clubs was complimentary but drained both time and energy. The days of the "free fieldman," who had time to attend night meetings across six states, were numbered. Yet, it would not be until the 1970's that the Holstein field staff would make a complete break from this tradition of public relations, and start providing practical advice to breeders on a fee-for-service basis.

Recognizing Brood Cows

While there had been recognition programs for sires for years, there was no program for the recognition of the great brood cows of the breed. It wasn't until 1957 that the distaff side was recognized. This delay was primarily due to numbers — the limited progeny from a given dam made it difficult to develop a fair way to recognize superiority. But the Board, after deliberation, presented a Gold Medal Dam award proposal to the 1957 convention that required that at least three offspring qualify for type and production standards. The program was popular from the start, with 69 cows qualifying the first year.

From the start, the Gold Medal Dam program's main objective has been to recognize genetically superior cows on the basis of proven genetics from several progeny. Breeders who have worked with Gold Medal Dams have often used the program as a launching pad for their breeding and merchandising efforts. Since 1957, over 3500 dams have qualified for the coveted award.

Like the other Association recognition programs, the Gold Medal Dam award requirements have been updated periodically. The most recent change, in 1984, bases the award entirely on the performance of the dam's offspring.

Fobes Mechthilde Ollie (left) and D. H. Posch Agnes Elaine were two of the first group of Gold Medal Dams selected in September 1957.

Glen M. Householder

Finishing The Fabulous Fifties

After long years of service, several valuable leaders retired after serving the breed and the Association most of their lives. Hod Norton, after ably chairing the Research Committee for five years, retired in 1958. A year later, A. Horace Winchester concluded a continuous record of 41 years of service with the Association. Retiring the same year was Glen M. Householder, who had served as Extension Director from 1938 to 1955 and as Special Assistant after that. The longevity of these leaders is an example of the dedicated work of hundreds of men and women who have had long careers of service with the Association. It is a tradition that still remains.

In one respect, the Association ended the fifties the way in which it started —with the headquarters office crowded and cramped. But the situation had been anticipated and land for expansion had been purchased as it became available in the neighborhood. The Association crossed into the new decade with a new expansion project underway.

The fabulous fifties were a transition period for the Holstein-Friesian Association, both in leadership and in growth of the breed. The organization established itself as the majority dairy breed in the country, a leader among its peer herd book associations. A network of state and local clubs had flourished and a national force of professional fieldmen and classifiers was in place. The membership was growing, operations were becoming more efficient, and the financial position was strong. Things were looking good as the 1960's approached.

Broadening the Genetic Base

Happy 75th birthday! The 1960's began with the celebration of the Association's Diamond Jubilee at the annual convention in Syracuse. Holstein leaders from all over the world gave the meeting an international flavor and, in a taste of global things to come, initiated the World Friesian Conferences. Convention delegates voted to revamp the entire membership structure, limiting memberships (except for existing life memberships) to a 10 year period. Meanwhile, back in Brattleboro, the second expansion to the office, adding another 50 percent in space, had been completed for less than the original cost estimates and paid for from operating revenue. In a tribute to the "pay-as-you-go" attitude of the Association, reserve funds had not been touched for either building addition.

And 1961 was another milestone year, the 100th anniversary of the first traceable Holstein imports to America. As if to symbolize the progress of the Holstein breed, the nation had its first 300,000-pound producer, Zeldenrust Pontiac Korndyke 2649438 (EX). When she died the following year at age 16, she had produced 306,051 pounds of milk and 11,649 pounds of fat. (Her record was broken in 1965 by College Ormsby Burke 3420439.) In 1965, another grand matron of the breed passed away at age 21. Minnow Creek Eden Delight 2494802 (EX) owned by Meadowfarm, Orange, VA had a lifetime record of 12,211 pounds of fat, the highest level to that date of butterfat production ever recorded in the Association's files. The Holstein industry, with the Holstein Association leading the way, was finding new genetic tools to help produce even better animals.

Executive Secretary Rumler called the era "The Soaring Sixties." And soar did the Association! While the number of dairy cows in the United States continued

Under Robert Rumler's leadership, the Holstein Association reached out to serve more breeders during the 1960's.

to drop each year, Holstein registrations kept growing. But there was no room for complacency, no time to rest on laurels. Those with vision knew that to continue to stay ahead, the Association needed to improve programs and reach out to embrace and serve a larger segment of the Holstein industry.

The theme of the 1961 Convention was: "BROADENING THE GENETIC BASE - WITH REGISTERED HOLSTEINS," a slogan that charted the course for the Association in the 1960's. The period was, in a classic sense, a struggle between traditionalism and progressive ideas within the Holstein industry. The annual meetings of the period reflected the ongoing conflict. But the flying sparks at board meetings and annual conventions showed the dairy industry that the Association was truly a membership organization, and while it might sometimes take years to resolve an issue, each member had a voice through the delegate process, and the delegates were not hesitant to speak out on the issues.

In his 1961 address to the annual meeting, Secretary Rumler told the delegates: "The strength of a breed is reflected in the strength of its commercial following." He then cautioned them not to be content to trade on the fact that the purebred registered cow is accepted by all as the animal of superior merit needed to improve the breed. Rumler called for the members to tell their story with conviction and to recognize the need to develop a broader genetic basis in the industry, saying: "There must be a greater number of animals of known ancestry from which to choose in order to provide the basis for advancement in the years to come."

The Association began to shed the cloak of "elitism" that surrounded registered breeders and opened up a populist movement that would continue through the 1970's into the 1980's. Association President Fred Nutter raised the issue at the 1963 convention when he asked, "Should we cut our cloth to fit only the few top breeders, or should we aim toward attracting 15 or 20 percent or more of the better dairymen interested in Holsteins?" Continuing, he said, "...Right now we seem to have reached a plateau....It is time to ask ourselves some soul-searching questions." He asked the delegates whether they were going to be content to register merely 5 to 6 percent of the Holsteins in the country, or, "Since it is generally recognized that practically all improvements in dairy cattle have come through the breeders of registered stock, could we not effect greater improvements by bringing more into the fold?"

The controversy over whether to expand registration to include red and white as well as "off-color" animals came to a head in the 1960's. But the heat and smoke generated by the color marking controversy tends to obscure the real developments of the period: the major advancements that were achieved in breed improvement techniques. Not only was the complete Holstein production base brought into the Association computers, but at the same time, a new descriptive classification program clearly defined and measured, for the first time, economically important type traits. By the decade's end, the whole concept of evaluating animals had begun to change. As the attitudes of the members toward traditional color and markings softened, the genetic base of the breed was broadened by the addition of animals which previously could not have been registered, high quality performers with genetic strengths that surpassed their markings and color.

Instead of milk production records and classification scores alone, the industry was beginning to use genetic measures. The color of an animal's hair (red & white or black & white) or the location of the color spots and patterns was no longer as important. With the Soaring Sixties had come the realization that there was much more to genetic advancement than meets the eye.

The Computer Age Arrives

Mechanization of the Holstein Association not only improved services to members by speeding registry and other paperwork processing, it also unleashed the power of data processing on breed improvement programs. By 1970, data stored on tape in Brattleboro would change the way breeding decisions were made.

Registry completed its first full year of partial mechanization with the IBM equipment in 1960. In the same year, a committee chaired by Murray Wigsten, with DeWitt Mallary and Leon Piquet, investigated the further automation of all Association operations. Their report, a long-range plan for further mechanization, was approved by the Board of Directors in December 1961. And, expressing

By 1969, the Association had the production and type information needed to issue performance pedigrees in its upgraded computers.

continued optimism and support for the growth of the Association, the Board unanimously approved a five-year expenditure of $350,000 for automation.

The plan called for the installation of an IBM 1401 tape system, the most advanced at that time. A Performance File would be created, consisting of type information on all classified animals as well as the production records for all cows tested under official breed programs during the preceding 30 years. The task of transferring records was protracted: each typewritten record had to be recorded on punch cards before it could be transferred to tape, but, by the end of 1964, two months after the new computer arrived (and installed in a special room with controlled humidity and temperature and a reliable power source), two million punch cards had been transferred to tape, with many yet to go. The new mechanization precipitated a major revamping of the rules for naming animals; prefixes for each herd were required for computer operations, effective January 1, 1965.

And these improvements were only the beginning. The Performance File had information readily available which took days to find in the Type and Production Year Book (Green Book). By the end of 1966, the Association was on a 10-day inquiry cycle and had 11 people in data processing and 53 in registry. By mid-1967, the performance file inquiry was reduced to a seven-day cycle and with another equipment upgrade (to an IBM 360/30), the first computer printing of registration certificates began.

The new computer capacity allowed the storage and processing of valuable breed improvement data which could be extracted to guide breeders in their decision-making. A major overhaul of the Performance File was started to add sire summaries for type. By the end of the decade, the Association, now having all official production and type information needed to go with the lineage information, began once again to issue pedigrees. The Association had, by wise use of computers during their infancy, combined type and production information to build the only complete data bank of ancestry information on U.S. Holsteins, a unique and exclusive position that has been retained to this day.

Maintaining Integrity

Integrity and reliability, the hallmarks of Association records since the beginning, were further strengthened in the 1960's by several new practices. First, a formal "Code of Show Ring Ethics" was adopted in 1966 by the Association, and was added to the already-established codes for the conduct of public sales as well as the handling of export affairs with registered Holsteins. (The show ring code would be again strengthened in 1984.)

Second, delegates at the 1966 Wichita convention voted to change the bylaws to allow the Association to publish the names and addresses of non-members with whom the Association would not do business due to infractions. Up until now, Holstein Association bylaws had called for the publication in the Secretary's Report of cases in which members had been brought before the Executive Committee and penalized for conduct unbecoming a member of the Association. While this was an effective disciplinary tool, it did not protect Holstein Association members from fraudulent non-members. The new change enhanced the overall public announcement policy, a disciplinary policy that has remained an effective tool to protect the integrity of Association records.

As the decade began, the Association had said farewell to one of the men who had helped build the Association's reputation. Clinton M. Horn, who had done so much, starting with the 1919 Cabana case, to uphold the Association's integrity, retired in 1961 at age 75. He had succeeded Judge O. U. Kellogg as General Counsel in 1935. He was followed in this important post by Robert B. Atkinson.

The Harrisburg Association Disbands

The genealogical base of the Association was expanded in the late 1960's in an unusual way — by the absorption of a rival, but small herd book operation, called the Holstein-Friesian Registry Association, Inc. of Harrisburg, PA, commonly known as the Harrisburg Association. Spawned by dissidents after the Cabana case and the bitter proxy fight of 1921, the organization was formed by Dr. Howard Reynolds in 1925. Small in size and operating on a limited budget, it was strictly a registry organization, offering no other services to members. Because of

"...it is what's underneath the hide that counts."
—Maurice S. Prescott

this, it could offer low fees and therefore attracted quite a lot of business for several years, enough, in the words of Secretary Hod Norton, "to be a nuisance and serious annoyance to the Holstein-Friesian Association of America and its membership."Yet, during the period when tuberculosis-eradication was a major dairy health problem, the Harrisburg Association had provided an important service to the breed since the Commonwealth of Pennsylvania and other states recognized their registry papers in order to qualify reacting animals for purebred indemnities.

After Dr. Reynold's death in 1942, his assistant, Ruth Watkins, who had learned the business while working in Brattleboro for Secretary F. L. Houghton, tenaciously kept the organization going for over 20 more years. Faced with dwindling membership, the Harrisburg group voted in 1966 to discontinue operations and accept the Holstein Association's offer to take over the records. In September, all Harrisburg records were shipped to the Brattleboro office and by the end of 1967, 12,892 animals had been re-registered, and 154 former Harrisburg members had applied and been accepted into the Holstein Association ranks. A long period of frustration and confusion had ended. There was, once again, one Holstein herd book.

Struggle Against Traditionalism — The Color Controversy

M. S. Prescott, writing in the *Holstein World* about the 25-year battle over color markings that had come to a head in the 1960's, said in a 1967 editorial, "It is generally recognized that it is what's underneath the hide that counts." In a historic decision, the delegates to the 1969 annual meeting agreed. Meeting in Anaheim, CA, they amended the Association bylaws to allow the registration of purebred Red and White Holsteins (in a separate Red and White herd book) and the registration of "off color" purebred females in the regular herd book. It was the first major change in color requirements since 1910.

The first Holstein herd book published in 1872 had no detailed color marking requirements. Likewise, when the Holstein-Friesian Association was formed in 1885, there were no specific color markings in the bylaws other than a single note that animals must be black and white to be registered. In 1920, the Executive

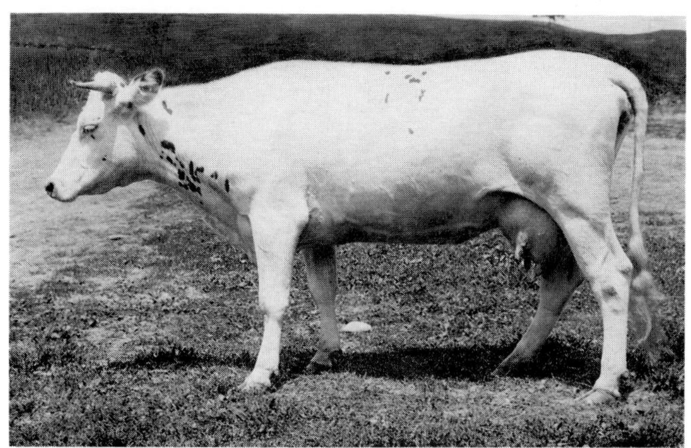

Light-colored DeKol 2d was admired by Holstein breeders in the early 1900's.

The influence of dark animals like Sir Pietertje Ormsby Mercedes first brought the color issue to the 1944 convention.

Committee, wanting to outlaw any animals that looked as if they might have other than Holstein ancestry, drew up a set of specific color marking requirements that stood, virtually unchanged, for six decades. These approved color markings were pleasing to many breeders and, in the eyes of some, had a definite trademark value to the breed.

In the early 1900's, light-colored animals brought the highest prices and were most popular with breeders. This popularity was most likely due to admiration for DeKol 2d, one of the prolific foundation cows of the breed. Then, from 1920 to 1940, animals which were predominantly black were favored due to the influence of the Ormsbys, particularly Sir Pietertje Ormsby Mercedes and his progeny. Other great dark animals such as Dunloggin Woodmaster, Marathon Bess Burke, and Montvic Rag Apple Marksman advanced the popularity of the black animals, and then the problems began. Mating these mostly black animals generation after generation, breeders started to get an occasional animal with black running down the leg to the hoof, or with black running low in the tail or the switch — color markings which barred registration.

Secretary Norton, writing in the 1942 annual report, first warned members, "It seems like a very fine point of discrimination when we record one animal with black running down the leg to within a half inch or perhaps only a quarter inch of the hoof and we reject another when the black runs down to the hoof..."

The color issue was first raised at the 1944 annual meeting in Columbus, OH,

The approved color markings, unchanged from 1910 to 1969, were pleasing to many breeders.

and the resolution was defeated on a voice vote. It was here that one of the oft-quoted remarks was made by the renowned breeder, Mark Keeney, who said in exasperation of the lengthy debate over black running down into the hoof or into the switch, "Haven't we more useful things we could discuss? We're not breeding switches." Delegates to meetings for the next 20 years would echo his sentiments, but would not have the votes to change things.

The controversy continued throughout the 1950's, fanned by an article in late 1951 in the *Holstein World* by Dr. E. S. Harrison called "Milk or Switches." A special committee on color markings made a report to the 1952 Roanoke convention, and again, the resolution liberalizing markings was soundly defeated. Over a decade passed before the subject was brought up again, this time at the 1966 Wichita meeting. The resolution was debated in open forum but was later withdrawn, leading to the appointment of a committee chaired by Harold Shaw, consisting of George Lorenz, Max K. Herzog, M. S. Prescott, and Dr. J. L. Albright. The committee presented a report in 1967 that not only recommended liberalized color markings, but called for Red & White Holsteins to be registered. Yet, in spite of a proposal of "provisional" registrations of such "off-color" animals, the color issue was again defeated at the 1967 annual meeting in Minneapolis.

But, as happened with other issues over the years, it took outside pressures to break through the barrier of traditionalism. In this case, the Holstein-Friesian Association of Canada already had a committee hard at work on color markings. A U.S. group of Chairman Lawrence Caldwell, Dr. L. O. Gilmore, Eugene Nelson, and Donald Seipt worked with the Canadians and brought forth recommendations that, after hours of debate at the 1969 Association convention, were finally adopted.

The new color rules required that so-called "off-color" Holstein females, born after July 1, 1969, could be registered as long as the suffix "OC" was included in their name. Red and White Holsteins of proven registered ancestry would be registered in a separate herd book effective January 1, 1970. (The separate herd book requirement was removed a year later.)

Future conventions would further broaden the rules so that by 1984, color was

After hours of debate, the color marking requirements were relaxed at the 1969 convention, 25 years after first being considered.

no longer a major issue. In 1976 the restriction barring registration of first generation offspring from "OC" animals, when the progeny did not meet color marking rules, was lifted. Off-color bulls could likewise be registered, with the "OC" suffix, as of July 1, 1976. The delegates to the 1984 St. Louis convention, on a vote of 116 to 102, passed a bylaw change eliminating the "OC" designation. Now, as long as an animal had proper ancestry and was not all white, all black, or all red, it was eligible for registration. The color issue, which had absorbed so much energy and emotion, was, indeed, history.

Advances In Evaluating Type — Descriptive Classification

The classification program of the Holstein Association was initially set up to help breeders receive recognition for their cattle. "Taking the show ring to the farm" was an expression used in the early days. The system was designed to rate animals according to their relative merit and their "percentage of perfection." But over the years, the classification program took on a different perspective. It no longer was just a recognition program, but a merchandising tool, and eventually, a breeding tool.

Descriptive classification gave Holstein breeders a better tool to use in making AI breeding selections.

Dairymen had recognized the value of herd classification since World War II and the program had continued its steady postwar growth into the 1960's. In the period from 1952 through 1967, the number of Holsteins being classified each year quadrupled.

Year	Registered Holsteins	Year	Registered Holsteins
1952	16,963	1960	55,906
1953	20,429	1961	59,186
1954	23,024	1962	64,986
1955	26,107	1963	70,124
1956	30,208	1964	74,123
1957	38,803	1965	72,691
1958	44,899	1966	81,535
1959	57,004	1967	84,052

In December of 1965, Jack Fairchild, a long-time classifier, was appointed Field Coordinator of the classification program to observe the day-to-day work of classifiers. In September 1967, Maurice Mix took over as Director of Classification Services, supervising the field and office operations of the program until 1978. During this period, the classification field staff grew from six full-time and three part-time, to 23 full-time classifiers.

"It takes a special person to be a classifier," Mix explained during a 1984 interview. "Not only must he have the ability to score cattle correctly and uniformly with his peers, but he must also be able to communicate with breeders. We look for people with a unique cow sense, people who have milked cows, who know the business. There's often a lot of bucks riding on a final classification score. It takes a real pro to be a good classifier."

And the Association knew this. Its leaders had gone to full-time classifiers in the fifties and now were holding type evaluation conferences three to four times a year to maintain a high degree of competence and uniformity among inspectors. But the system was getting outdated; it was time for a change. The change would

Avery Stafford is a good example of the axiom: "It takes a real pro to be a good classifier".

become known as Descriptive Classification.

Descriptive Classification evolved in the mid-1960's, starting when the 1961 annual convention voted to have a special committee take a look at the herd classification program. A group headed by Fred Nutter made a report to the 1962 annual meeting, and after feedback from delegates, conducted further study and prepared a report for the 1963 annual meeting. Then, in 1965, Dr. George Trimberger, a nationally known judge and educator, was appointed to a temporary position on the Association staff to evaluate the classification program and suggest improvements. Trimberger, an expert in the field of dairy cow type and body conformation, took a leave of absence from Cornell University to do the work. With the assistance of key Association staff members, he developed a more precise method of recognizing functional type traits of animals, assigning descriptive codes (1 through 5) to indicate strengths or weaknesses in terms of the classifier's concept of "what is ideal." The system described 14 functional type traits that were considered important to clearly describe the animal.

"We wanted to paint a picture of the daughters of a bull." That is how Secretary Rumler described the change to Descriptive Classification. It was to be the first basic change to the type evaluation system since 1929.

Animals would continue to be rated and receive a final score based on the four major uniform scorecard items: General Appearance (30 points), Dairy Character (20 points), Body Capacity (20 points) and Mammary System (30 points). The final scores were categorized:

EXCELLENT 90 to 100	GOOD 75 to 79
VERY GOOD 85 to 89	FAIR 65 to 74
GOOD PLUS 80 to 84	POOR 64 or less

Dr. George Trimberger points out functional type traits to a Cornell group.

But now, in addition to final score, there was a practical system to describe the animal's functional type traits. It was the first effort by a breed organization to graphically define type measurements in practical, readily visualized terms — to give a "word picture" of the animal's appearance. The Descriptive Type Classification Program went into effect January 1, 1967.

The new program proved to be popular and would not be replaced until 1983. Since its introduction, more than four million registered Holsteins have been evaluated in the United States as well as thousands of others overseas. Classification for type became a major breeding tool and the Association, with a staff of 35 classifiers in the U.S. to evaluate herds at home and abroad, classified five times as many cows at the end of the 16-year life of Descriptive Classification than when it began:

DESCRIPTIVE CLASSIFICATION

Year	Registered Holsteins	Year	Registered Holsteins
1967	84,052	1976	269,573
1968	88,534	1977	290,135
1969	87,215	1978	321,315
1970	99,361	1979	358,685
1971	131,769	1980	413,023
1972	133,461	1981	429,657
1973	144,523	1982	466,600
1974	175,409	1983	444,660
1975	254,563		

Not only was the program popular, it also reemphasized the importance of practical, functional type. Breeders began to give greater attention to conformation than at any time before, emphasizing "wearability" — the ability to produce and reproduce consistently and efficiently for several lactations. While the final score was still important to many, other breeders sought help in pinpointing strong and weak points in animals and selecting sires to improve specific trait weaknesses. Descriptive Classification gave them the tools to do it.

The Field Force — A Period Of Transition

The mid-sixties were a time of transition in the field. The national staff, now called program directors instead of fieldmen, had grown to 13 but the turnover rate was high. There was not much meat and too much potato in the program director's work, and morale sagged.

From the start of the field program, fieldmen had been promotors of the breed, providing information to breeders at the request of the breeders. They worked with the state and local associations and covered many miles attending meetings and shows. But there was something missing in the work — the "nuts and bolts" contact with breeders. Something else was also missing.

Zane Akins was hired as program director for Wisconsin in 1966, later rising through the ranks to Executive Secretary. He explained another problem the field staff had: "Keep in mind that many of us had come off the farm and had owned cattle and were intensely interested in the breed

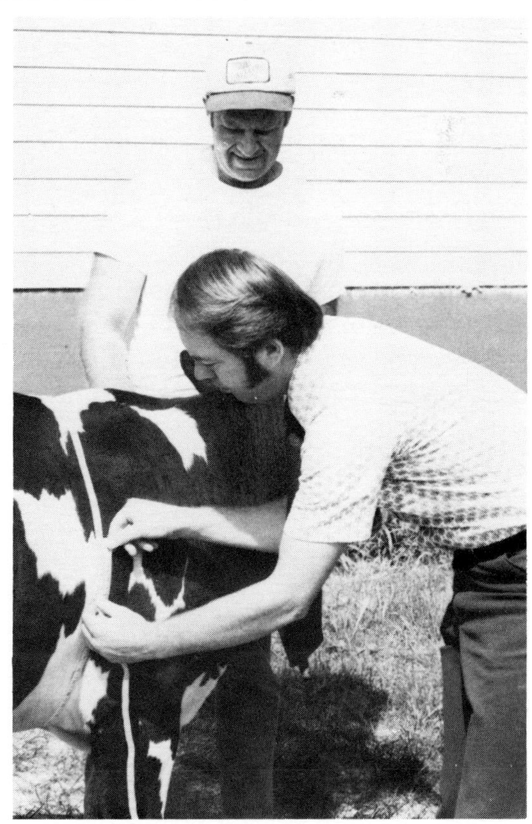

Zane Akins began his Holstein Association career as a program director in Wisconsin.

Charles Howe typifies the experience and expertise that Holstein fieldmen bring to the job.

and its progress. Once we became members of the staff, because of the potential for a conflict of interest, we could not 'join in' the excitement of ownership." A policy change in the 1970's would loosen that prohibition.

But what about the satisfaction of dealing one-on-one with the professional breeder? The "free fieldman" was an Association tradition. Holstein program directors were not supposed to be able to walk onto a farm and advise a breeder on sire selection or help arrange a purchase or sale, even though, since they made the rounds and knew their stuff, they were always being asked.

Economics dictated a change. Field work was an expensive component of Holstein Association operations and, as the inflation of the 1960's put pressures on the budget, leaders looked at Field Services for new sources of income. This resulted in a pivotal move by the Association in the early seventies — to a more commercially oriented field program. Now, when program directors performed individual services for a breeder, there would be a charge attached to the service.

As might be expected, this change was not received enthusiastically at first. Program directors, in many cases, had been doing some of the work for free: spotting cattle and passing along leads, giving informal advice on sires, and helping put people in touch with one another for sales. Now they were to charge for these services. The reaction to the new situation calmed as the program evolved and breeders realized the value of the service being provided.

Once the direct service programs got rolling in the 1970's, the stimulation of being able to provide individual technical assistance added new energy and enthusiasm to the field force. The willingness of the Holstein Association Board to go from a free field service to a fee-for-service system had turned out to be an important swing point in Association history. Generating new revenues, it helped keep the registration fees lower. This "full-service" concept would blossom and flourish in the 1970's.

Developments In Production Testing

The sixties were a time of change in production testing — not particularly in the methods, but rather in the manner in which the testing information would be used to support a new method of production proofs for sires. (See next chapter.) The

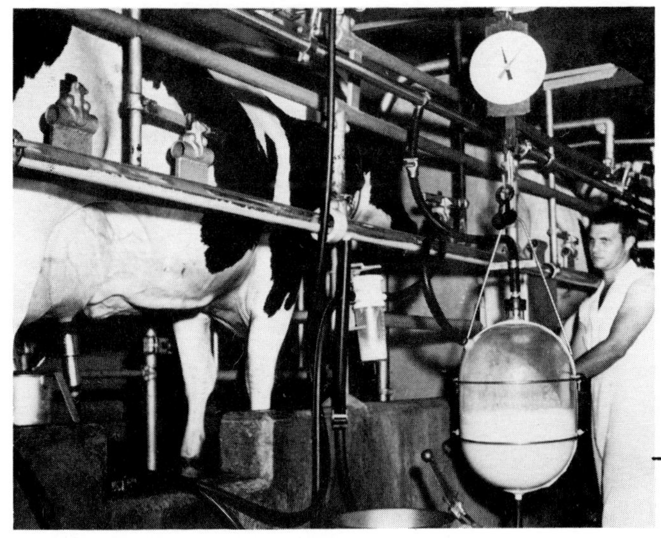

DHIR became the sole production testing program of the Association in the late 1960's.

decade also saw the start of a testing program for solids-not-fat, which had, at its peak, 25,000 cattle enrolled. A little ahead of its time, the program faded, yet it was the first effort by a breed organization to record nutrient components of milk in addition to butterfat. Also during the decade, two of the initial production testing programs, Advanced Registry (AR) testing and Herd Improvement Registry (HIR) testing, both of which had been faltering, were terminated.

The Holstein Association's DHIA Acceptance program was adopted by all breeds in 1960 under the label of Dairy Herd Improvement Registry (DHIR). DHIR took the standard DHIA records and testing program and applied special breed association rules and requirements.

Some breeders, used to the credibility of the HIR and AR programs of the Association, were suspicious at first of the supervision and accuracy of the DHIA programs which varied from area to area among the more than 1400 local DHI associations. As early as 1940, the Holstein Association had taken an active role in standardizing production testing through the Purebred Dairy Cattle Association (PDCA), as Secretary Hod Norton chaired the first committee on Uniform Rules for Official Testing. Over the ensuing years, PDCA continued its oversight role, particularly of the breed association's production testing programs. As DHIR grew and concerns in the purebred industry continued, PDCA used its experience with testing rules and procedures to help strengthen the DHIA programs.

The Association worked with other breed organizations to write the special rules and procedures to protect the credibility of the DHIR records. The Association also lent its support to strengthening both the DHIA and DHIR programs, meeting in 1964 with the individuals in charge, along with Holstein breeders, in the 10 largest dairy states. In April, 1965, USDA announced the formation of the National Dairy Herd Improvement Coordinating Group. Bringing together 13 dairy industry leaders, including Secretary Rumler representing PDCA, the group's mission was to strengthen the DHIA program. This was, in the eyes of some, the first real effort to provide a single overall

Dr. Dean Plowman

approach to the national DHIA program. The Coordinating Group, tackling issues such as uniform bylaws and testing procedures, has done much to raise the credibility of the DHIA programs with Holstein breeders. The DHIR program has continued to grow.

AR testing, which had been the backbone of production records since the days of Solomon Hoxie, had outlived its usefulness — each year the number of participants decreased. By 1964 only 59 cows were enrolled. It had served a vital and important role in breed improvement and promotion over the years, but now breeders needed information on all the animals in the herd rather than on a few selected animals. The program was discontinued in July 1965.

Herd Improvement Registry (HIR), the Association's herd testing program for 30 years, began a decline in usage in 1959 as the DHIA programs became dominant. The development of DHIA Computing Centers hurt the HIR program as many breeders sought the lower cost DHIA program. Each year the number of herds on HIR dropped dramatically. Due to this fading interest, the Board of Directors discontinued HIR in 1967.

DHIR, which added special enhancements to the DHIA testing program, included only the registered animals in DHIR member herds and gave complete lactation period results for the tested animal regardless of the length of lactation, as well as providing 305 and 365 day and lifetime records. By 1968 it was the sole production testing program of the Holstein Association.

New Breed Improvement Tools — Sire Summaries & Pedigrees

Computer-based production testing records opened up a whole new realm of possibilities for breed improvement. In what may turn out to be one of the truly important developments in Holstein Association history, the Association, after lengthy and difficult negotiations with USDA, led the way in establishing an acceptable industry-wide plan and program to accept daughter-herdmate comparisons for sire proving, sire recognition, and sire evaluation for production. In doing so, the Association entered into an agreement with the USDA's Agricultural Research Service (ARS) to receive all DHIA production records on Registered Holsteins and USDA-computed sire summaries. This

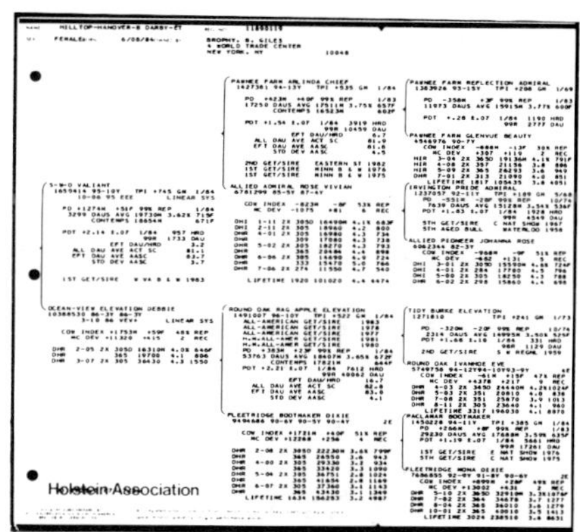

Computer-produced pedigrees were first issued in 1969 by the Holstein Association.

pioneering role further enhanced the stature of the Association as the leader in the U.S. dairy industry.

The new herdmate comparison system, developed by Dr. C. R. Henderson of Cornell University, had been adopted by USDA to bring a new level of credibility to the sire production proofs. The formula for breeding value for production had been developed by Dr. Dean Plowman of ARS who, using the method of predicted differences, designed a system that measured all sires by the same standards. No longer were non-AI bulls segregated from AI bulls. As Dr. E. L. Corley, Chief of the Dairy Cattle Research Branch of USDA at the time said, "By expressing breeding values on all bulls in terms of predicted differences, it is now feasible to evaluate and directly compare all bulls included in the summary as to their transmitting ability for production."

It was time to abandon the daughter/dam comparison method of sire evaluation — it had been recognized as being obsolete. Dr. H. J. Schmidt, speaking to the 1967 convention, said, "The fact that a very high percentage of our daughter-dam proven sires has failed to meet production promise after entrance into AI is evidence that we need to re-evaluate. This failure does not mean we did not have the bulls or cannot breed them; rather, that our selection and interpretation procedure has lacked effectiveness."

After lengthy negotiations, an agreement was reached with the Agricultural Research Service. It called for USDA to provide the Holstein Association with the registered Holstein Sire Summaries based on herdmate comparisons three times a year. The Holstein Association had led the way for other dairy breed organizations. A key element in the agreement, one which Executive Secretary Rumler pressed for, was that the Association would receive production information from all registered Holsteins on standard DHIA test (400,000 lactation records a year at the time). Here was the information the Association needed to produce performance pedigrees. When combined with Descriptive Classification data, it meant that the only complete production and type data base on Holsteins would reside in the Association's computer tapes in Brattleboro.

By 1969, using the information gained from USDA, as well as the type data already in the computer, the Holstein Association was back in the pedigree business. The computer-produced performance pedigrees, unique among breed organizations, utilized information from DHI as well as DHIR records, show ring winnings, production leader lists, individual classification and cow index records, and the most recent sire summary data on both type and production. The information was gathered, organized, screened, and printed automatically. The days of searching laboriously through the "Green Book" for such information were about over.

Genetic Advances In The Late Sixties

The production testing and type information was now in a system where it could be readily used by breeders, and it was. The results show it.

The transition from phenotypic to genotypic breed improvement began in the late 1960's. Phenotypic measurement — what we see and measure — had worked for years. Dairymen, drawing from their experience, had formed genetic conclusions from two or three daughters of a cow, not recognizing that it was too small a sample to be accurate. Since the phenotypic system had gotten the breed to where it was in the industry, it was not easy to discount countless show ribbons and test records pinned up on the milkhouse wall. It was a topic that could, and still can, spark many a debate.

But after 1967, when breeders started using the new genetic tools, the results were dramatic. Genetic progress was showing up in the amount of milk in the bulk tank. As noted on the graphs in the Foreword, milk production improvement, resulting strictly from genetics, swung sharply upward in the late 1960's, increasing 150 pounds of milk each year. The work done by researchers, USDA, AI studs, Extension Service, DHIA, and the Holstein Association had come into place — and genetic progress was happening. The Holstein Association had soared through 10 years of declining dairy farm and dairy cow numbers with the strongest record in its history, reaching new highs in virtually every area of endeavor. It had broadened the genetic base of the U. S. Holstein breed and, with new genetic evaluation tools, was ready for the 1970's.

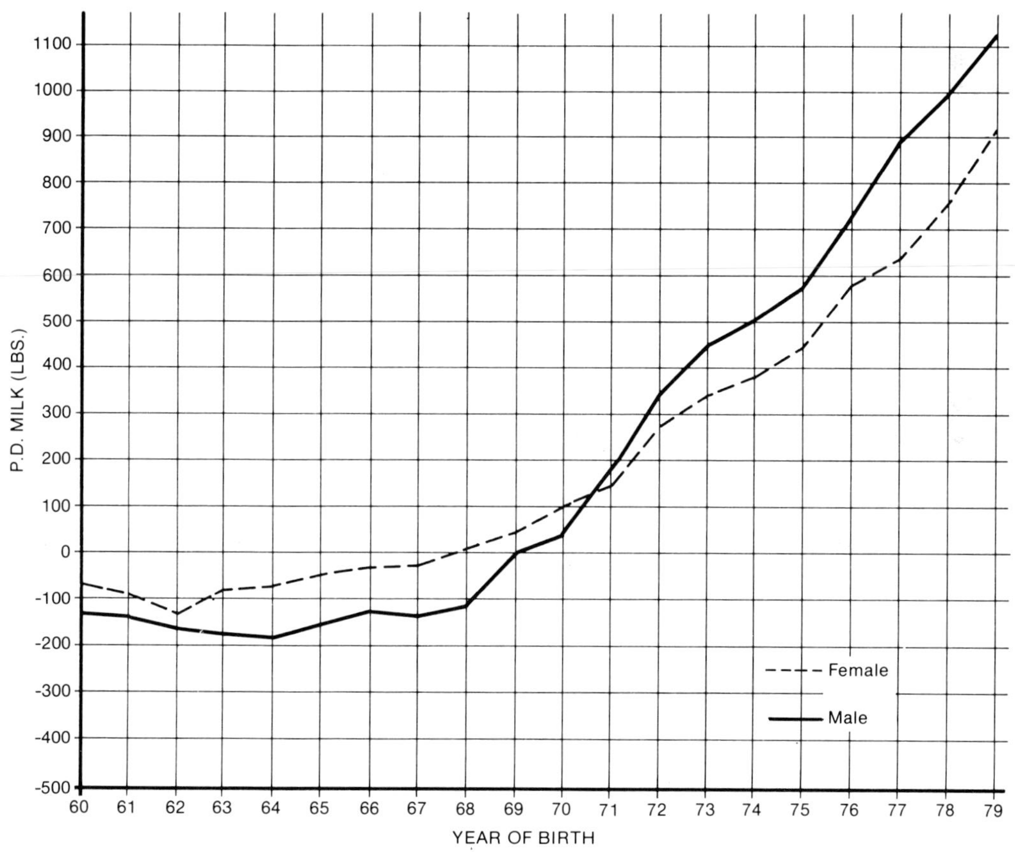

Genetic Trend for Milk Production.

chapter 5

Reaching Out to the World

The decade of the 1970's was a period of unparalleled growth, both at home and overseas. International marketing programs of the Holstein Association literally took off, as Holstein-laden DC-8s and 747s flew to countries all over the world. U. S. Holstein genetics — bulls, heifers, frozen semen, and later, frozen embryos — traveled the globe as foreign dairymen recognized what their American counterparts had already learned: that the U.S. Holstein dairy cow that had captured over 75 percent of the American market by the mid-seventies was a feed efficient, high producer that could easily adapt to changing diets and climates. The steady foreign market development by the Holstein Association had paid off, spawning a worldwide demand for U. S. Holsteins that started in the late 1960's and has continued into the 1980's.

As detailed in the next chapter, back home in the United States, the Holstein Association was reaching out with a professional field staff to offer an array of "full-service" programs aimed at not only their existing members, but also at the large commercial Holstein component of the breed. Grade Holsteins were accepted into the identified and classified ranks for the first time. New programs were launched under the flag of the new business corporation, Holstein-Friesian Services Inc., and in a major change in philosophy, the Association began charging for services being provided to individual breeder/members by the field staff. And the era saw Robert Rumler, one of the most forceful and forward-thinking leaders the Association ever had, pass the day-to-day Executive Secretary management reins over and take on a more policy-oriented role as Executive Chairman.

Broadening The Horizons To The World Community

While the Holstein cow was becoming recognized as the premier dairy animal in the United States, it was also, through the efforts of the Holstein Association and individual exporters, gaining great popularity throughout the world. The global impact of U.S. Holstein genetics in the last 25 years is one of the remarkable stories in the history of the Holstein Association. While essentially all Holsteins in the United States and Canada traced back to the original 7757 imports, no Holsteins had been imported since 1905. The pendulum had swung — as much as 50 percent of the blood of dairy cattle in Europe now traces to North American Holsteins. Robert Rumler, writing in the Foreword to this history, calls it a process of "reverse osmosis."

However, it took many years to develop overseas markets. The Association had looked abroad for decades, starting with a trip by Secretary Houghton and W. S. Moscrip to Holland and England in 1925, where they reviewed the methods and progress in breeding. Yet, early exports consisted mainly of cattle that accompanied missionaries to foreign lands.

The first efforts of the Holstein Association to sell cattle abroad were directed toward countries in Central and South America. Europe was well-established with a dual purpose type of black & white animal and was in fact, along with Canada, chief competitor of the United States for the export market. In 1941, the Association's official judging manual was translated into Spanish and distributed throughout South America to vocational schools, agricultural libraries, and interested breeders. Shipments of Holstein cattle were made to Puerto Rico, Cuba, and Colombia until the outbreak of World War II brought the fledgling effort to a halt.

As the war ended, the Association established a Foreign Relations Committee, under the leadership of Owen Young, to begin the first formal export effort. Harvey Farrington was hired as the first export manager, and with help from R. E. "Rube" Everly, an experienced farm manager from Carnation, began the first attempts at foreign market development for the Association. During the period of 1946 to 1950, Farrington made a number of trips to Latin America, helping to open up South American acceptance of Holstein cattle. The only real success

Judging foreign cattle shows was an effective export marketing tool in the postwar years. Here Glen Householder prepares to leave on a trip to judge at the Peru National Holstein Show (May, 1957).

came when two herd books were established. In 1944, A. C. Oosterhuis traveled to Colombia to help organize a pedigree record organization — Associacion Colombiana de Holstein-Friesian, and several years later, S. B. Hall assisted the Holstein-Friesian of Ecuador in setting up their system, a herd book that has remained a strong herd book since.

Problems surfaced from these early attempts at international marketing, problems that would plague the industry for decades. American breeders were reluctant to quote on requests from South America, partly because of the complicated finances, but usually because they didn't have the documentation available on their cattle. Foreign buyers wanted photos, ancestry information, and extended pedigrees, and few farmers had the necessary paperwork on hand. The cattle boats sailed without their heifers.

Aside from making contacts and developing relationships with foreign breeders, Association representatives often served as judges for South American cattle shows. Even though the Foreign Relations Committee was terminated in 1948, judging, along with classification, continued and was one of the most effective export marketing tools. Throughout the early 1950's, export activity was minimal, being constrained by serious dollar exchange problems in South America since most countries, with the exception of Venezuela, had restricted currency relations with the United States — their money could not be converted into dollars.

Then two things happened that provided the catalyst to develop a strong export program for U.S. Holsteins: first was the recognition by the Holstein-Friesian leaders of the potential of the international marketplace, and next was the passage of Public Law 480, the Agricultural Trade Development and Assistance Act. Executive Secretary Rumler made exporting a primary objective of the Association and used the new program to help do it. The development of the Association as an international force in marketing the Holstein breed is, in the eyes of many, the most important legacy he has left the dairy industry.

"Part of the philosophy we had," Rumler said during a 1984 interview, "was to bring into the orbit of influence of our Holstein Association as much of the world dairy industry as possible."

The Buga, Colombia Holstein Show.

USDA/FAS Cooperator Program

Public Law 480 was set up, among other reasons, to address the currency exchange problem which was stifling agricultural exports. Procedures were established to allow countries with "soft currency," currency not convertible to dollars, to pay for U.S. agricultural exports with their own money which was then usually returned to them in foreign aid or as payment for their goods.

Shortly after passage, market development work was added to the FAS program and the Holstein Association jumped at the opportunity. In 1956, the Holstein Association Board approved a six-point plan for international market development. It involved:

1. Cooperation with the Foreign Agricultural Service (FAS)
2. Direct assistance to and intensified public relations with Latin American countries
3. Spanish language publications
4. Advertisements and news releases
5. The establishment of a "Code of Ethics" for Holstein export agents
6. Other activities pertaining to the expansion of foreign markets

The FAS market development program, designed to stimulate agricultural exports, was a creative coupling of private sector and government funds, with both groups paying their share of the activity. Under the regulations of Public Law 480, certain businesses and trade organizations, willing and capable to expand U.S. agricultural exports, were selected to be cooperators. The Holstein-Friesian Association has been designated a cooperator since the early days of the Act, having been the first livestock organization to participate. It has remained one of the most active cooperators for over 20 years.

The Holstein Association signed an agreement with the Foreign Agricultural Service of USDA to promote the sale of U.S. Holsteins overseas. Matching the "market development funds" from FAS, the Association began a systematic program of promoting the export of Holstein genetics, a program that has had far-reaching consequences. The FAS sponsored projects involved a great deal of foreign travel by key Holstein-Friesian Association executives, and also included participation in international livestock trade shows, providing cattle judges for

Mr. & Mrs. Rumler at the 1959 London Dairy Show.

overseas Holstein shows, and sending classifiers to evaluate foreign Holsteins. The development work also was accomplished through advertisements in foreign trade journals and by teaching foreign dairymen at educational seminars. The long trips and continuous contacts with foreign officials paid off. Association activity related to the FAS Cooperator Agreement was largely responsible for stimulating a strong export market for U.S. Holstein breeders.

International Export Development

Once FAS funds started flowing in earnest, Holstein Association officials hit the road to promote their dairy animals. Executive Secretary Rumler took a six week trip to South America in 1957 where he visited all Central American countries as well as Ecuador, Peru, and Venezuela. Europe, however, remained an impenetrable market. In 1959 Rumler and his wife Jean attended the 50th anniversary of the British Friesian Society as Association leaders started an attempt to make inroads into the European dairy market. The first major penetration came in Italy where Carnation Farms, which had a representative stationed abroad, shipped Holsteins to a large dairy near Rome. And while Italy was the first market success in Europe, the real breakthrough came after the Pabst dispersal sale when a U.S. sire, Pabst Ideal, was purchased by Dr. Gunther Rath for a fledgling AI center near Hanover in West Germany. It was a shipment that pierced the armor of resistance fo U.S. Holsteins because the bull did extremely well and paved the way toward acceptance of Holsteins from the United States, (Chapter 8).

Exports of Holsteins started to climb dramatically in the mid-sixties. In 1964, a Holstein calf was featured in full page ads in the *Wall Street Journal* and leading financial magazines. The ads touted the fact that TWA was shipping jetloads of calves to Italy — and Italy was where the Holstein Association, in 1966, sponsored the first overseas show and sale of registered Holsteins. Forty-five head of cattle, from all over the U.S., were flown to the International Dairy Show in Cremona where they were a great success, prompting several visits to the Association offices by foreign breeders as well as a return engagement the following year.

Holsteins from all over the U.S. were flown to the International Dairy Show in Cremona in 1966, sparking great enthusiasm for the breed in Italy.

As interest in Holsteins grew, so did the problems caused by the complexity of international cattle dealing. Holstein Association leaders were learning that international sales require the ability to adapt to complex and ever-changing situations. The continuous change of leaders in some countries dampened any attempt to build a long-range program. Holstein specialists learned to search for a sparkplug who could ignite a program — in some countries it was a private breeder; in others, it was persons in the government. Robert Rumler explained it like this, "Each country had its pocket of leadership — some were considered radical at the time. Breed development in a country depends on so many things — but it takes some good, solid, constructive leadership, from whatever source that might be, private or public."

Additionally, there was a need in the marketplace for quality control over the registered Holsteins being selected and shipped abroad. The complex problems being encountered called for new linkages to procure, check, and ship the cattle. The leaders of the Holstein Association decided that there was one way to get the job done properly, and so, in the late sixties, they created a new business corporation: Holstein-Friesian Services, Inc.

Holstein-Friesian Services, Inc.

Holstein-Friesian Services, Incorporated, (HFS) is a taxable business corporation wholly owned by the Holstein-Friesian Association. Operating at "arm's length" from its parent corporation, it markets Holstein genetics — cattle, embryos, semen, and technical assistance — both internationally and domestically. Through 1983, HFS had generated over $70 million in Holstein sales since its beginning in 1969.

The 83rd annual meeting of the Association took a pioneering step in the marketing of registered Holsteins when delegates, meeting in Milwaukee, voted on June 26th, 1968, to authorize the formation of a new subsidiary corporation, to be known as Holstein-Friesian Services. HFS was designed to provide services not available through regular Holstein-Friesian Association channels, thereby expanding the Association's capability to serve its individual breeder-members. During its 16-year history, the corporation has received criticism along with praise.

HFS was created, among other reasons, to fill a void in the international cattle marketing process, to tackle a problem area that had been encountered since the 1940's. The problem came to a head in the mid-sixties when the Association conducted an intense campaign to promote the merits of U.S. Holstein cattle, an effort which generated a great deal of interest in the breed from foreign buyers. But, as before, one of the major constraints to selling Holsteins overseas was the inability of the Association to follow up on potential orders. Too often, after developing a client in a foreign country, Association representatives would have to watch the transaction flounder and be lost because of problems of potential exporters. Sometimes, the problem was the inability to finance the transaction or a lack of familiarity with the intricacies of international sales, but often it was simply a shortage of capable exporters (fewer than 10 firms were active in the business). In some cases, foreign government officials and financing groups preferred, or were required, to deal with a representative organization with the reputation of the Association instead of individual export firms. HFS was set up to overcome these trade barriers in order to help the owners of U. S. registered Holsteins sell their cattle overseas.

Holstein-Friesian Services was incorporated in Massachusetts in November, 1968, through an issuance of 100 shares of no par value stock to the Holstein-Friesian Association. Initial capitalization was $30,000 with the sole stockholders being the HFAA Board. HFS was to be based at the Brattleboro home office and had these officers: Chairman of the Board R. DeWitt Mallary; President Robert H. Rumler; Executive Vice President and General Manager Charles J. Larson.

The Board of HFS is composed of nine directors. By agreement, the Association President, Vice President, Executive Secretary, and Chairmen of the Finance and Executive Committees serve with four other elected members of the Association Board. The Board President of the Association serves as President of HFS.

Management positions in HFS are filled as follows:

The Executive Secretary of the Association serves as Executive Vice President and Clerk of HFS

The Association Treasurer serves as HFS Treasurer

HFS Director and staff work for the HFS Executive Vice President

The first initiative of the new corporation was outlined in the 1968 HFAA annual report: "...expansion of export marketing activity was determined by the Board of Directors to be the most logical for initial development." The plan was to stimulate maximum foreign demand and then help members market their animals to meet that demand. It worked.

This new approach to overseas marketing blazed a remarkable trail for Holstein exporting. By 1973, HFS was shipping over 10,000 animals a year to countries all over the world. These efforts opened the door, both with contacts and procedures, for others. Now, over 60 private firms compete in the U.S. Holstein export market with Holstein-Friesian Services, Inc., and many prominent exporters of U. S. Holsteins gained their experience and their international contacts working with or for HFS.

During the first year of operation, HFS launched a two-pronged approach to international exporting. C. T. Barns, hired as the first director in March 1969, traveled extensively to coordinate the promotion of Holsteins at fairs in Italy, Spain, France, Portugal, Brazil, Venezuela, and Mexico. In addition to the promotional activities, HFS tackled and solved a traditional problem area — finding quality animals to promptly fill an export order. Franchised export coordinators, each capable of taking responsibility for a shipment from farm to dock, were set up in a nationwide procurement network. HFS helped to coordinate the shipping, obtain insurance, procure the animals, select an export facility, and combine orders to make up a cost-effective load.

The Original
HFS Export Coordinators
1970

Duane Green
Alvin Piper
Donald Larkin
Lowell Knief
Murray Wigsten

One of the key ingredients to the early success of HFS was the quality control, particularly from a genetic standpoint, that the organization provided its customers. Not content to just sell cattle, the corporation used the resources of the Association to provide technical assistance and applied technology to foreign buyers. Holstein-Friesian Services, from its first months, set standards of operations that are trusted throughout the world today. This "marketing integrity as well as Holsteins" approach to exporting was one of the first, and one of the enduring, achievements of the organization.

1970 marked the first full year of operation. The highlight of the year was the shipment of 82 bred heifers to Yugoslavia, the first such export of Holsteins to that country. In order to facilitate the shipment of pedigreed Holsteins, HFS combined loads with other breeds as well as grade Holsteins. Two planeloads of mixed livestock went to Greece during the year while other shipments of Holsteins went to 16 other countries.

This exporting success continued the following year when sales volume grew to over $3 million. In June, 1971, over 500 registered Holstein and Hereford heifers

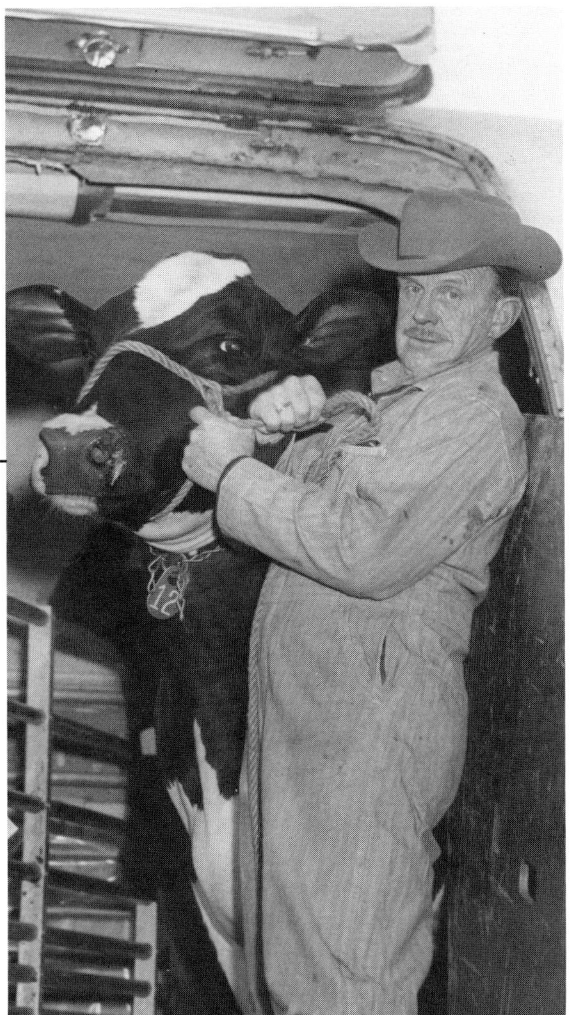

Dick Brooks helps load one of the 157 animals that were shipped to Italy in 1973.

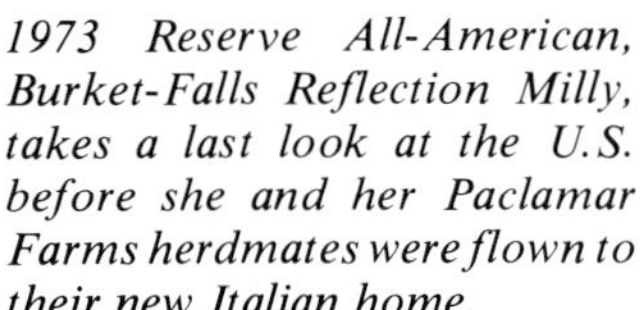

1973 Reserve All-American, Burket-Falls Reflection Milly, takes a last look at the U.S. before she and her Paclamar Farms herdmates were flown to their new Italian home.

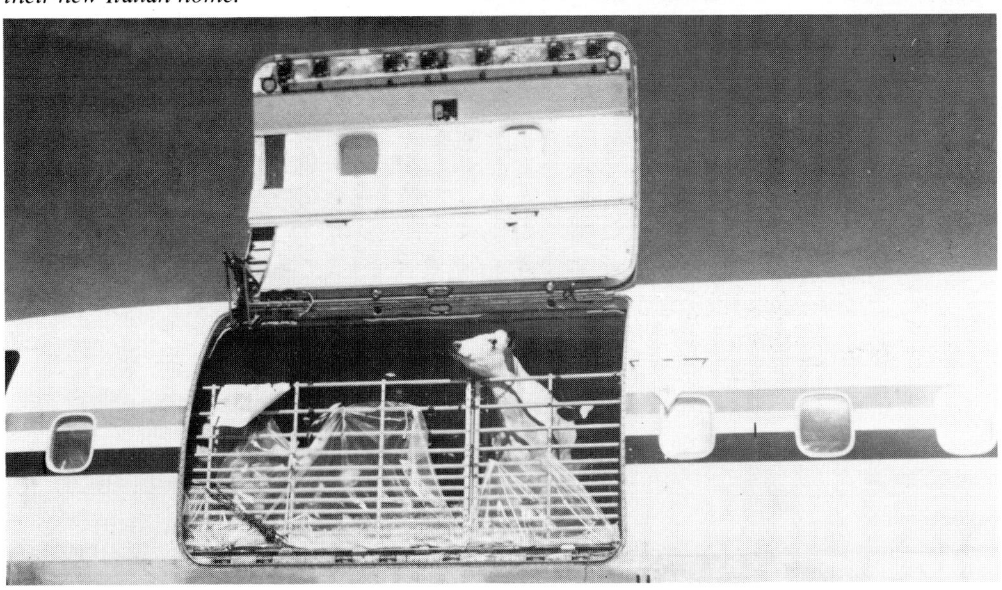

Some heifers left home reluctantly but once overseas, adapted well to their new surroundings.

Large shipments of U.S. Holsteins left by ship under HFS contracts in the early 1970's.

HFS received the President's "E" award in 1975.

left Richmond, VA, by ship for Italy and Spain under an HFS contract. The month before, five cargo jets of heifers had departed for Europe, making this the largest combined air and sea shipment ever of U.S. registered cattle. HFS broke even financially for the first time.

The export activities of the business corporation continued to grow. In February, 1972, 606 Holsteins, including the renowned Paclamar Farms registered Holstein herd from Louisville, CO, were exported to Italy. This was the largest private international sale of an entire Holstein herd to a foreign buyer and therefore attracted international attention. A record 10,260 animals, of which 8512 were registered and grade Holsteins, were shipped during the year. By 1973 HFS was exporting a thousand cows a month, needing one ship and at least one aircraft a month to do it. The number of Holstein breeders supplying cattle had increased as well: one HFS shipload of heifers came from 14 states and 246 individual breeders.

The high export volume continued in 1974 when over 9000 animals, including 6443 Holsteins, were sold to 20 countries. The corporation, after paying an $83,000 fee to the Association for management and office services, realized a substantial profit solely from export sales.

At the 1975 annual meeting, Association staff noted that while HFS had 16 percent of the Holstein overseas market, competition was growing. The success of HFS was drawing new exporters into the marketplace. Officials in Washington noted the achievements as well: in September, HFS was singled out for national recognition, receiving a presidential award, the President's "E" Award for Excellence in Exporting. Awarded from the Department of Commerce, the "E" was presented to Chairman of the HFS Board Gordon W. Newton, President Robert H. Rumler, and Vice President Charles J. Larson by Secretary of Agriculture Earl L. Butz. Other staff and the full board were also present for the presentation. This award capped a highly successful export era, a period that soon was to change for all U.S. exporters, including HFS.

Holstein-Friesian Services suffered a series of setbacks in the latter half of the seventies. While there were many factors involved, the primary cause was the skyrocketing costs of air transportation resulting from the OPEC oil embargo.

Other factors that affected profits included strong domestic competition, a drought in Europe, and the inflated prices of U.S. cattle. HFS export sales dropped dramatically (see Table A) as a result of these problems.

Table A
HFS International Revenue

1969	$ 113,567
1970	270,358
1971	731,660
1972	1,970,953
1973	8,460,590
1974	8,648,121
1975	8,862,620
1976	7,512,575
1977	8,843,328
1978	5,883,563
1979	2,854,999
1980	2,848,754

In 1976, HFS learned the lesson that international marketing is a high risk business. An intricate international joint venture corporation (IFRAM) had been formed involving HFS and a French corporation, the Agricultural Development Bank of Iran, and several other Iranian companies. A major financial blow was struck by the government of Iran when it reneged on the payment of shipping costs for five planeloads of cattle already delivered. HFS had been supplying cattle to Iran since 1973 and had shipped over 2500 head. In December 1975, HFS discontinued further shipments to Iran because of the default of $342,000. For the first time, HFS had extended credit for air transportation, and suffered the consequences. In spite of diplomatic and civil efforts, including a direct appeal to the Shah of Iran, the corporation was never paid and suffered an actual loss of $234,000.

There were several bright spots during this period of adversity. HFS opened up

An HFS shipment of registered Holsteins is bound for South Korea.

new markets for Holsteins in Korea and Russia for the first time and, in 1977, put together the largest single Holstein export order — more than 5000 heifers to Hungary.

However, when several key employees left to set up their own export business, the corporation was left short-staffed in a critical period. Soaring domestic cattle prices and the economic problems caused by the embargo of grain to Russia, factors which hurt the overall export market, triggered a moderate loss in 1978 and several years of major financial losses. HFS was not the only concern facing adversity: total dairy cattle exports dropped from the 1975 high of 73,277 to less than 13,000 in 1980. During that same period, HFS shipments fell from more than 7500 head a year to just 832 in 1980.

The most significant achievement of HFS was the new international markets that it opened for registered U.S. Holsteins, and in doing so, shored up the value of Holstein cattle on farms across the country. In 1972, Association President A. C. "Whitie" Thomson, recognizing this accomplishment by HFS, estimated that the export success of the new organization had added $100 to the value of every registered Holstein in the U.S. With 1.5 million living registered Holsteins, he calculated the effort had added $150 million net worth to the collective balance sheet of the owners of Holsteins. And considering a lesser, but still important, impact on the value of grade Holsteins, another $225 million could be added. Thus, for an initial investment of $30,000, it could be argued that the international trade that HFS spawned added nearly $400 million to the net worth of Holstein dairymen.

Specially designed crates carried eight to nine Holstein yearlings to Yugoslavia.

International Achievements

The Holstein Association presided over an unprecedented expansion in foreign markets for U.S. Holstein genetics in the 1960's and 1970's. "Jet-set" Holstein heifers were whisked off to countries across the world and major dairy industries were developed overseas using U.S. Holsteins. The vision of Owen Young, Bob Rumler, and others who pushed the export business was paying off.

Moving only slightly faster than their forebears who had traveled from Europe by sailing ship, the cattle exported during the initial years traveled by "Slow Boat to China." Losses of cattle during these early shipments were not uncommon, usually the result of human ignorance of the temperature requirements of the animals or failure of the ventilation systems. As ships became air-conditioned and able to maintain the temperature below 65 degrees F, surface shipments became a safe economical alternative, and are still used in the 1980's.

The advent of the jet transport changed the exporting of cattle dramatically — now in less than 24 hours, cattle could be delivered virtually anywhere in the world. And they were. At first, Boeing 707s and Douglas DC-8s were used, later the freighter versions of the jumbo jet Boeing 747 were common. A typical load on a 747 is 180 bred Holstein heifers, each weighing about 1050 pounds. Holstein Association specialists who have flown on the freighters report that the cattle are docile and arrive in good shape. David Goolsby, a Far East sales manager for HFS noted, "In the old days when we shipped to the Far East, the trip took 12 to 16 days. Even with plenty of food and water aboard, we'd expect up to a two percent loss. With air, we haven't lost any animals."

Holstein Association ads in foreign agricultural publications have been an effective marketing technique.

Modern transportation equipment made Holstein exporting more visible and helped spur interest in the breed. Agricultural publications as well as newspapers and magazines featured black and white cows arriving in foreign climes. The Holstein Association generated many of these news articles as a market development activity. As the Association's international program evolved, it became a total marketing package, stressing the need to buy genetic quality in the animals and semen being purchased as well as the need for technical assistance to take advantage of the high quality animals being acquired. In providing a complete program, the Association offered countries or areas of countries an evaluation of the potential for establishing and developing a Holstein breed based on U.S. genetics. After the initial contacts and research by Holstein specialists, a comprehensive plan for improving the dairy industry could be developed, and then a detailed selection of animals and semen could be made over a period of time to make the plan work. One excellent example of the success of such a market development program occurred in Hungary, where Holstein Association expertise completely changed the country's dairy industry (Chapter 7).

But, while publicity was effective and Holstein experts, working under the FAS agreement, were expanding their contacts across the world, the growth of exports of U.S. Holsteins was stimulated by the same things that had made the cow so popular across this nation — efficiency and adaptability. The Holstein was proven in many tests to be the most efficient converter of energy among dairy

U.S. Holsteins have flourished in northern climes ...

cattle. In a 1974 study, J. K. Oldenbrock of Holland compared Holsteins from North America with Dutch Friesians and Dutch Red and White cattle. He found that in order to produce 100,000 kilograms (220,000 pounds) of milk a year, it took 19 Dutch Red and Whites, or 18 Dutch Friesians, but only 14 Holsteins. The same study showed that the European cattle required over 63 kilograms of Total Digestible Nutrients (TDN); the Holsteins required only 58.7 kilograms.

Another strong advantage of the Holstein breed in foreign markets was the breed's adaptability to environmental factors and its success at producing milk, lots of milk, on many types of feed. This had been proven to some extent in the United States where Holsteins had flourished on small hillside farms in Vermont as well as 1000-cow dairies in Florida, Arizona, and California. When that adaptability was demonstrated in foreign climes, it helped spread the Holstein's popularity all the more.

and adapted well to tropical settings.

And, as in the United States, there were changes in the international marketplace on which the Holstein Association was able to capitalize. In Europe, the dairy industry had, since the turn of the century, continued to develop the Friesian as a dual purpose animal — for meat production and milk production. The European animal was smaller, more compact, and had a slightly higher dressing percentage but as people's preferences changed toward leaner meat, the marketing door was opened for the larger, higher producing U.S. Holsteins.

When these marketing factors were exploited by the Holstein Association in the mid-seventies, the export of Holsteins skyrocketed. As previously noted, the trigger for this increased export activity was the formation of Holstein-Friesian Services, Inc. Finally, there was an organization to help take care of this pent-up demand for Holstein cattle. Holstein exports accounted for over 90 percent of the dairy cattle shipped from the United States during the 1970's. The activity, which reached a peak in the middle of the decade, can be estimated from the records of cattle that were inspected by the Animal and Plant Inspection Service (APHIS) for export.

Table B
U.S. Holstein Exports
Grade And Registered

Year	Exports
1971	11,949
1972	17,145
1973	31,750
1974	54,292
1975	67,273
1976	52,083
1977	68,098
1978	48,241
1979	29,245

Mexico purchased thousands of Holsteins, tapering off toward the end of the decade as the peso was devalued. As noted in the HFS section, both Hungary and

Iran were leading importers of Holsteins. The Republic of Korea had been the leading Asian importer, having made a strong commitment to building a dairy industry based on U.S. Holsteins. Japan continued to be the most consistent importer, as it had been in earlier years, of a limited number of genetically high quality animals each year. As the decade ended, Holstein executives were scouting new territories as Executive Chairman Rumler traveled frequently to the Soviet Union, to Cuba, and as part of the first official agricultural mission to China.

The Holstein Association had achieved its goal — to expand the breed to the world community — and had seen Holstein genetics become a part of the dairy herds of over 60 countries. But while international marketing was an important new initiative, it did not detract from the primary responsibility of the Association — developing, improving, and expanding the Holstein breed in the United States.

Many high quality U.S. Holsteins have been exported to Japan.

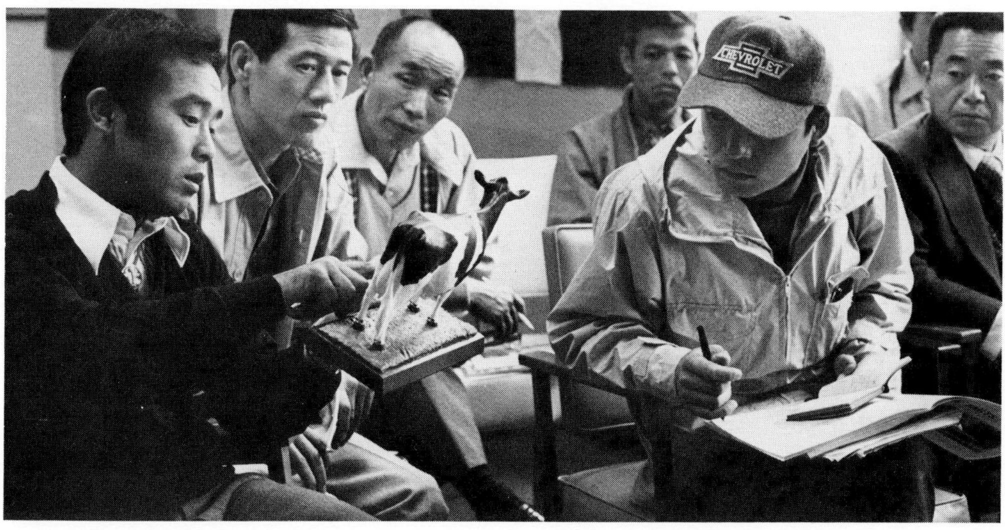

A Full Service Organization

The Association's theme for the 1970's was "Full Service" as the decade started with momentum from the 1960's. Association leaders developed a long-range plan, the third such 10-year plan during Secretary Rumler's tenure, and in a change of emphasis, called for providing better service to individual members. This would change the role of the field force, a change that had been developing since the late sixties. The objective was to accomplish improvements to the breed on a herd-by-herd basis by instituting a new "on-farm" approach. With the expanded genetic base and new evaluation tools ready to put to work, the Association was able to launch a series of new services. Ray H. Kliewer, research and development director, spent the early seventies on the full-time development of new programs and services, as well as the review and overhaul of old ones. The result was a full array of field services which would be delivered under the banner of the commercial arm of the Association, Holstein-Friesian Services, Inc.

Several steps were taken during the period to open up the Association, to better communicate with members, to become less of a "country club" and more of a broadly-based membership corporation. Open forums and general discussion sessions at annual conventions were expanded. Winter forums, where delegates to the forthcoming annual convention and members had a chance to discuss upcoming convention issues and provide input into the Association programs, began in 1974. These Board-led regional sessions proved to be an effective way for the leaders of the Association to relate to the members and obtain feedback from them. As the decade ended, the Association began market research and opinion surveys to find how it could better serve its members.

This GEM logo was developed for use by GEM clients and for office use on GEM forms and ads.

Genetic Evaluation And Management Service

The first new field program of the 1970's was a herd improvement program, Genetic Evaluation and Management (GEM), which was started in 1971. GEM was designed to give personalized consultation and advice on breeding and was operated by the Association's business arm, HFS, through a network of franchised consultants. These private entrepreneurs, working on a free-lance basis, also helped dairymen with record-keeping as well as the marketing and purchase of cattle. Al Bates, Ray Tanner, and Walter McClure were the first persons franchised by HFS. By the end of 1973, 18 consultants working in 21 states, had evaluated 17,099 animals in 273 herds. At the 1975 annual meeting of the Association, James Pound, the Association's Director of Extension, described the practical assistance being provided by the GEM program. Pointing out the advantages to the Association, he noted, "...Dairymen will continue to relate positively to this organization and the registration process if services provided contribute in a practical way to their success and an improved standard of living for their families." The on-the-farm advice from GEM consultants was doing just that.

One of the more successful of these franchised people was Clarence Stauffer in Ephrata, PA, who quickly gained the respect of his peers in the Pennsylvania Dutch farming communities and showed how well the GEM program could work. He later would be hired as a Holstein consultant on a full-time basis.

The Association had decided to charge for all individualized services being provided by the field staff to individual dairymen and now wanted to utilize their own staff. Holstein Association program directors had been shedding some of their traditional public relations roles and were working directly with dairymen to introduce GEM service and to help members with foreign marketing. In the mid-seventies the Board of Directors, realizing that transfer and registration fees could no longer support "free fieldmen," officially shifted the field emphasis from public relations and promotion to one of direct assistance to farmers. The GEM program was transferred from franchised consultants to Association program directors and the 18 Holstein Association specialists began to charge for the genetic, management, and marketing help they provided. During 1977, they

Jim Pound headed up the Association's GEM program.

assisted more than 500 herds.

During the early years the GEM program aroused concerns from the sire procurement and mating personnel from the AI studs already at work in the field. It was not unlike the friction caused by the Association's fledgling HSDS program (next section). While relationships were unsettled at first, the long-term results have been good for the breed. The Association's James Pound, who dealt directly with the factions as the GEM project got underway, explains, "At the grass roots level, the interfacing of Holstein staff with the AI staff, created by the need to become better acquainted with the top bulls and their breeding patterns, ultimately created a better climate than would have existed without the GEM program."

The field programs of the Association continued to expand in the mid-seventies as the work responsibilities of the program directors (formerly fieldmen) were redefined. Supported by specialists like Charles "Chuck" Detch from the home office, the field staff began to limit their traditional public relations roles and concentrate on providing direct technical assistance to dairymen. A new program, Herd Builder, which provided dairymen follow-up breeding advice after the classification of a herd, was added to the field service programs in 1977. Over time, the herd improvement program evolved into what is

Beecher Arlinda Ellen set a world's record for milk production in 1975 with 55,661 pounds.

now known as Corrective Breeding. In January 1983, it was transferred from HFS to the Holstein Association Field Service Division.

Holstein Sire Development Service

Executive Secretary Rumler, writing in the *Holstein-Friesian World* in 1970, noted that a better system of young sire development was needed: "Only about 250 young registered Holstein bulls enter AI studs each year. There is a critical need for sampling many more carefully selected young bulls. Our breed has a rich reservoir of potentially young sires."

And so, in 1973 one of the more far-reaching genetic advancement programs of the Holstein Association, known as the Holstein Sire Development Service (HSDS), was begun under Holstein-Friesian Services, Inc.

A survey had been conducted in the early seventies querying the private owners of bulls that had a +400 PDM. The results revealed that most owners had become discouraged in trying to test their bulls and, in fact, had slaughtered them before the proofs were completed. Breeders did not have confidence in the genetic tools of the breed and were reluctant to take the risk of sampling young bulls. In conclusion, the survey revealed that too few young sires were being sampled for optimum genetic progress. At the time, fewer than 900 bulls were being sampled by AI centers and private breeders, and it was estimated that three times that number should be sampled...if possible. So the need was there for a progeny testing program, and, while members indicated that they would support such a program, there was widespread concern from the AI industry. The Board of Directors and management took a deliberate approach to entering sire development assistance, and still, HSDS was a controversial program at the start.

Other breed organizations had developed sire testing programs but, due to the majority position of the Holstein Association, the concept of HSDS created a stir in the AI industry. There was concern that the Association intended to set up its own AI stud and semen distribution system. The concept was fought vigorously at NAAB conventions and through an organized effort to convince the Association Board to drop the idea. Some seemed convinced that the Holstein Association would bias the proof of their test sires, especially with the type

HSDS has opened up new opportunities for breeders to prove young sires.

proofs, by classifying them too high. It was a hot issue, but, in September 1973 after deliberations at several meetings, the Board approved HSDS.

With a gale of controversy swirling, Zane Akins was chosen to set the sails for this new service. Basically a one-man operation at the start, Akins found himself making the rounds to talk to owners, screening bulls, and overseeing the sampling and the distribution, all the while trying to allay the fears of AI by personal contacts. He recalls one crucial meeting: "I remember sitting with Doc (Harold) Schmidt, then the manager of Carnation Genetics. He offered me a cigar and I smoked it. I think it was the first cigar I'd smoked in my life, but I won a friend and eventually Carnation became a cooperator by agreeing to take some initial bulls. Some might have called it a 'peace pipe.' "

HSDS was designed to help dairy farmers with young bulls who were having problems obtaining enough daughters to develop an official performance summary or getting daughters into enough herds to develop an acceptable confidence level. The program gave breeders the choice of retaining ownership of their young bulls during the sampling period so that the owners could benefit from any outstanding performance proofs that might be forthcoming.

During the first year, 21 breeders participated. By 1974, 174 "Qualified Sires" had been accepted, and by mid-1975, over 70 sires had been sampled, qualified, and sold for their owners by HFS. In order to be accepted into the program, the bull needed to have a pedigree that met basic requirements and be inspected to meet conformation and other standards. As the HSDS program developed, it grew from a simple semen distribution system to a total progeny testing program.

Success was not long in coming. At the Winter Forum in Ohio in 1974, a group of young breeders, who had a young bull ready for evaluation, aggressively questioned Akins, the Association representative, about HSDS. They argued that the program would not work. Challenged to give the program a try, they

The first success in the HSDS program came with Straight-Pine Elevation Pete.

enrolled the bull, one of the first six in the program, and he subsequently turned out to be one of the better in the breed — Straight-Pine Elevation Pete. Not only did his proof hold up but it actually improved as time went on. This success story, coming as it did so early in the program's development, was a major hurdle cleared by HSDS on the road to acceptance and respectability.

By 1984, when a young sire is designated "Qualified," 500 units of his semen are distributed to cooperating dairymen by Holstein Association field consultants. The plan is for each young bull to have calves in 50 herds and thus have a confidence level of 50-60 percent with his first proof. Since its 1973 beginning, HSDS has sampled 194 sires that met the entrance standards and has developed their performance proofs on a multi-herd basis. Like the GEM program, HSDS was brought under the Field Services operation of the Holstein-Friesian Association during the restructuring of HFS in 1983.

HSDS benefited the Holstein breed in several ways. Not only did it provide an option for the owners of good young bulls, it also prompted the AI industry to increase the number of young bulls they sampled. It opened up new opportunities for breeders: there have been 20 HSDS sires leased or purchased by AI firms in the United States and three overseas. And importantly, under the leadership of Akins (then James L. Copper, and presently, Dennis Funk), HSDS has found its niche in the overall sire proving system. The concerns of 1973 are history as the Association enjoys an excellent rapport with AI, with AI's one-time arch rival, Akins, at the Association helm.

State Organizations — Termination Of The Allotment System

The lack of a uniform extension effort continued to hinder the field programs of the Association. The time had come to revamp the 1958 agreement with the Ohio, New York, and Pennsylvania Holstein organizations because it just was not working well. Association leaders wanted operational control of extension activities in those states by using national program directors, while the states wanted the financial support from the Association to come with no strings attached. It had been a sensitive subject for years, but would finally be resolved in the 1970's and would become an important part of the Association's transition to

The Pennsylvania Holstein Association has been an active exporter of U.S. Holsteins.

a more business-oriented breed organization.

The first of the three states to acquiesce to the overtures of the Association was Ohio, which had suffered financial setbacks in the 1950's and 1960's. Extension Committee member James Lewis reported to the June 1968 Holstein Association Board meeting that the Ohio Holstein Association had arranged to accept a national fieldman.

The following year, Earl Noel and George Bridenbaugh, President and Vice President respectively of the Pennsylvania Holstein Association, and Avery Stafford and Leonard Baird, who held similar posts with the New York Holstein Association, met with the Association's Extension Committee to review fieldwork in the two states. The major area of concern was the limits placed upon the use of funds as well as the requirement of annual approval of budgets of the state associations by the national Association. The state representatives still wanted the allotment of funds from the Association, but wanted operational autonomy.

After considerable discussion, a new policy on the deployment of program directors was developed, a policy which recognized the autonomy of state associations on local issues, yet supported the states through continuation of incentives to state members from the national Association. The allotment system was to be phased out by January 1, 1971 and national fieldmen assigned to the two states. While it would take a number of years for the pot to quit boiling, a long-term controversy had ended.

The wounds from the allotment wars have healed and most breeders agree that the relationship between state and national is stronger than ever. State associations have played an important part in the history of Holstein fieldwork; in the early days state fieldmen were the only Holstein extension effort. And many dynamic leaders carried the Holstein torch to many a night meeting across the rural countryside. Two early standouts were I.D. Hadley and George Nichols in Ohio who used innovative measures to keep the state association solvent during the Depression years.

Art Nesbitt, Secretary-Fieldman in Pennsylvania, got people in his state accustomed to holding their county meetings whenever his date book was open on a given night. As his friend, Association staffer Jim Pound says, "There was never much thought given to the need for the meeting, what the program would be, except that Art would be there and he would be the program and if he came, why, they didn't have to worry about anything." Nesbitt's enthusiasm and energy were a driving force in Holstein field work in the 1950's in that key dairy state.

Later, the Pennsylvania Association, under the leadership of Bill Nichol, jumped right onto the export cattle boat with the Association when HFS was formed, emerging as a model of what state associations could do in the way of exporting, sales, and member support. And in more recent years, the Wisconsin Holstein Association has developed an entirely new posture with a strong board of directors and a capable state manager, Mike Snyder.

Holstein Classifier Don Cook (center) and Area Manager Bob Cain (right), review classification results with Tennessee dairyman, Larry Johns.

Improving Classification Data

The delegates to the 1970 annual convention approved a new reclassification concept and procedure which was implemented January 1, 1971. It required all eligible animals under five years of age in a herd to be classified without the inspector having prior knowledge of the previous classification. The desired result was to record changing type characteristics as animals matured.

The score and descriptive breakdown could be raised, lowered, or remain the same. History has proven this procedure to be one of the most progressive innovations in the era of Descriptive Classification. It contributed significantly to the accuracy of Sire Summaries and gave due recognition to the fact that animals may change for the better or worse as they develop.

More Tools For Genetic Advancement

When the herdmate comparison method of sire and dam evaluation replaced the daughter/dam method in the mid-sixties, it accelerated genetic progress in the Holstein breed. But, it was difficult for many dairymen to grasp the true meaning and value of "genetics," accustomed as they were to the individual considerations

New genetic improvement tools are explained by a Holstein Association consultant.

based on phenotypic accomplishments. Robert Rumler puts it like this: "We're talking about genetics now, we're talking about what the inheritance of the animal is, and what she's likely to produce — and the fact that she's Excellent-93 doesn't mean she's going to have all Excellent offspring or even a preponderance of them."

The dust from the 1967 Sire Summary system had hardly settled when the rapidly evolving genetic research prompted major changes. In the late sixties, a dairy magazine editorial had said, "we need to find out how good our bulls really are...not how good we can make others think they are." There was concern about the over-dependence on "hot" AI sires and whether the sire evaluation tools were adequate. A new evaluation system developed by USDA, Modified Contemporary Comparisons, came into place in the mid-seventies. Genetic evaluations were becoming more complex each year.

In 1975 the Holstein Association cut through some of the genetic haze surrounding sire and dam evaluation by putting both production and type information together in a way that, while complicated in its calculation, was easily understood and used by breeders. Researchers (Dr. John White, Dr. William Vinson, and J.A. Grantham) working for the Holstein Association at Virginia Polytechnic Institute confirmed a policy which had been adopted by the Association back in 1964, the need for a parallel breeding program, both for type and for production. Selecting sires for production would not necessarily result in genetic improvement in type traits or vice-versa. The research showed that both milk production and type must be emphasized, thus, the Total Performance Index (TPI) was developed. TPI combined the predicted milk production capability of a bull's future daughters with his predicted type transmitting ability. It went into effect in 1975 along with the system of publishing "transmitting profiles" of plus TPI bulls, which showed the influence of a bull on a given type trait. The TPI program was refined in 1979 when, in addition to selecting for type and milk, the dairyman could select for butterfat percentage as compared to the average for the breed.

An important genetic achievement toward the end of the decade was the adoption of the Best Linear Unbiased Prediction (BLUP) method of sire evaluation for type, developed by Dr. Hendersen and co-workers at Cornell. It is a sophisticated system that discards some of the biased information that resulted from the Herdmate Comparison, the system it replaced.

While in some cases it took a geneticist to completely understand some of the new genetic improvement tools, dairymen, with the technical assistance of

Holstein program directors, Extension Service personnel, and the AI industry, used the tools to continue the genetic advances begun 10 years before.

Sire Evaluation For Type (SET)

A fast, effective and reliable method of collecting type information on the daughters of a specific bull was developed by the Association and introduced in March 1976. A variation within the basic classification program, the SET program collects functional information as soon as a specified number of qualified daughters of a sire have freshened. If an owner, syndicate or group wants the type transmitting capability of a sire determined for a particular bull as soon as his registered or identified daughters freshen, and doesn't want to wait until the regular area program classifies the daughters, a special classification program can be arranged.

The SET program has proven to be a valuable way to prove young bulls for type traits and has been an asset to those who need to get functional type data as soon as possible. In the first year of operation, 19 applicants enrolled 42 bulls. By 1979, enrollment applications in the individual program had risen to 68 bulls a year and in the AI program, to 142 bulls a year.

Phil Holdaway, SET coordinator at American Breeders Service, in a 1979 interview, praised the program noting that, before its use, less than half of his bulls had type proofs. "We joined SET two years ago asking for classification data on first lactation animals," he said. "It really works well for us." In 1977 ABS enrolled 92 bulls in the SET program and had enrolled a total of 617 bulls through the end of 1983.

Since its inception, over 1800 bulls have been in the program and nearly all major AI firms in the United States use SET.

New "Ideal" Type Models

For over 50 years the "True Type" cow portrait had adorned the walls of Association offices, and "True Type" models had graced desks and mantels all over the world. The model had set a sense of direction for the industry for decades but, while excellent for its time, the 1922 version was a genetic antique — and not

"Holsti," unveiled in 1977, was the updated True-Type Ideal Mature Female Holstein model.

the angular Holstein of the 1970's. Leading breeders felt that the time had come to update the model and also to develop an example of an ideal young female. In 1973 the Board appointed a committee, under the direction of Richard Brooks, to work on a new model, a new Ideal Young Holstein Milking Female, starting with a portrait. Committee members were: Dr. Hilton Boynton, Dr. Harold Schmidt, Richard Keene, Thomas Lyon, and Director of Classification Services, Maurice Mix. The Ideal Young Female did not progress beyond the two-dimensional stage but the widely distributed reproductions of the portrait, which had been painted by Francis Eustis of Cincinnati, served well as a valuable type and breeding guide for the industry. The next task was to develop an ideal modern Mature Holstein Cow, a cow that is 57-58 inches at the withers, weighs about 1600 pounds, is 5 to 6 years of age, is in her third to fourth month of lactation, and is milking 120 to 130 pounds per day. The cow was one that would classify in the high Excellent category. The cow was designed to be walking slightly uphill, with the pin bones slightly lower than the hooks, with special attention given to make the cow more angular. A major change involved the design of the udder attachment, a high, wide udder attachment with the floor of the udder completely above the base of the hocks. The model cow, introduced in 1977, emphasized the "dairy-type" animal that had made the U.S. Holstein so popular across the world.

Donald Collins chaired the committee consisting of Eugene Nelson, Harmon Toone, Jack Fairchild, Elmer Dawdy, Richard Keene, and for the Association, Maurice Mix. The formal presentation of the painting of the Mature Female Holstein was made at the 1976 convention. In December of the following year, the Holstein Association Board approved the model of the Ideal Mature Milking Female.

The True-Type paintings and models had served the industry well for over half a century. The new designs reflected more accurately the phenotypic advances made over the years, representing the angular Holsteins that had gained such popularity across the country and the world. As the industry became more aware of the need to balance type with production, the new paintings and models gave a new generation of breeders a sense of the ideal concepts of conformation in female Holsteins.

Identified Holstein Female — Grade I.D. Program

The permanent identification of Holstein grade cattle was, like the color question, a lively topic of discussion in the 1960's and 1970's. Advocates on both sides of the issue argued their cases at three consecutive conventions while pressures mounted from outside, as well as within the Holstein Association, to put a program in place. Although the point was made that a high percentage of Holsteins of the 1970's were essentially purebred due to the extensive use of registered AI bulls, most Holstein Association members were in favor of maintaining a "pure" herd book. They, as well as management, stuck to their position and while other breeds have changed, the Holstein herd book is now the only "closed herd book" in the world. Identified Holstein grade females are recorded separately.

The first relaxation of attitudes surfaced at the 1971 Des Moines annual meeting when, in a surprise outcome, delegates voted 120 to 98 in favor of merging the Red and White herd book into the regular herd book. Members, many of whom had not actually seen a published herd book in years, affirmed that Red & White registered Holsteins were as pure, genetically, as their black and white sisters.

It was not so easy with grade Holsteins. Association members considered the issue at the 1973 convention and expressed "genuine concern" with the mechanics of the proposed grade identification program. They were especially worried about the cost of the program and how adequate identification would be assured. The convention voted to have the Board make a further study and report back.

There was a great deal of outside pressure on the Holstein Association to come up with a program. DHIA already had a grade identification program operating in several states. USDA was continuing to run into problems with the data used to prove sires since much of the production data lacked sire identity. There was a

The Identified Holstein female program went into effect July 1, 1976.

strong push from the academic community for an "open herd book." And the Canadians already had the National Identification Program in place. The die was cast. It was to be another struggle, like the off-color and red battles of the decade before, between traditionalists and a more progressive faction. The issue would last for several more years.

But the leadership of the Association was anxious to get an identification program going. With a full complement of service programs, they wanted to work these Association programs into the commercial herds, making the Holstein Association a more visible part of commercial dairying. So they brought a proposal, called Holstein Identification System (HIS), up for consideration at the 1974 convention and, after hearing lengthy testimony, watched the idea get tabled. During the discussion, Harris Wilcox, the well-known New York breeder and sale manager, brought about the only levity to the debate when he told of the farmer whose friends said that he must have seen a lot of changes in his 60 years of dairying. "Yes," the farmer said, "and I've been against every damn one of them."

Again, in 1975, the delegates voted the grade I.D. proposal down.

But the Board of Directors brought it up again at the 1976 Philadelphia convention and it passed. The Association's grade I.D. program went into effect July 1, 1976.

Within two years, over 50,000 grade Holsteins in 2200 herds had been identified. Dairymen were finding that grade identified Holsteins were worth more at sale time, an average of $100 per head. The grade I.D. program, which had been so long in coming, provided a needed service to hundreds of dairymen and, at the same time, by establishing a separate identification system, the Holstein Association herd book would remain a "closed" herd book.

The Rearick Case

One of the primary functions of the Holstein Association since 1885 has been to protect the integrity of the lineage of the Holstein breed. As Richard E. Nelson, who, as Special Assistant to the Executive Secretary has been instrumental in leading this endeavor, says, "Without credibility, what are the records worth?" Nelson, through pioneering work in the blood typing program as well as active

Skagvale Graceful Hattie set a new record for milk in 1971 with 44,019 pounds.

enforcement of Association rules and regulations, has done much to uphold this responsibility. While sometimes challenged, this ethical cornerstone, credibility, has been the foundation upon which the Holstein genetic data base has been built.

Yet, not since the Cabana case over a half century before, had there been a serious legal challenge to the authority of the Association over its membership until the Rearick case took place in the 1970's. Disciplinary proceedings against Allen and Sara Rearick, of Millheim, PA, started in early 1974 when the Association investigated the methods used to test their animals.

The issue was falsified milk production records. Rearick's production records had resulted in one national leader and six records in the top 35 on the National Honor List in 1973. The Rearicks were charged with deceiving the Association by misrepresenting the quantity of milk produced by animals in their herd. After a two-day hearing by the Association's Executive Committee, the production records for 1973 and 1974 were ordered to be stricken. But, like the Cabana case in the 1920's, the decision was challenged and a lengthy battle ensued.

In 1976 the Rearicks filed a suit in the United States District Court of Western District of Pennsylvania against the Association, alleging conspiracy among the breed and testing organizations to restrain trade. After two years of subpoenaing and examining records, gathering depositions and evidence, the decision was rendered. Judge Ziegler, on June 29, 1979, supported the Holstein Association in a six-page decision which was later published in the Federal Register. He affirmed the right of a trade (breed) association to discipline its members as long as the procedures followed were correct and lawful and as long as it acted in good faith, using due process. The five-year court battle upheld, in precise language, the principles established over 50 years before in the Cabana case.

Undesirable Recessives

Another method that the Association has used to uphold the integrity of the U.S. Holstein lineage has been to identify and deal with the issue of Undesirable Recessives. As the use of artificial insemination spread in the industry, and as the number of offspring from high merit bulls increased, it became even more important, for the genetic health of the breed, to gather information about any

Breezewood Patsy Bar Pontiac was the first cow of any breed to produce over 15,000 pounds of fat in a lifetime.

Mowry Prince Corinne was the first 50,000-pound cow.

inherited defects. Recognizing that such recessives could impact thousands of animals if not reported, the Association initiated a program of recording carriers of undesirable recessives in 1957. Owners of the cow that produced an abnormal calf would have the primary responsibility to report.

Next, in 1961 the Association issued a Statement of Policy which strongly discouraged the use of bulls that were recorded carriers of undesirable recessives and required appropriate identification any time the bull or semen was offered for sale. In 1977 the policy was expanded through a change in the bylaws, which called for the periodic publishing of a list of all carriers of undesirable recessives and, whenever the animal's name appears in a pedigree, it is to be identified with a code.

The abnormalities that have been defined as undesirable and transmitted through recessive genes are: Bulldog, Prolonged Gestation, Mulefoot, Hairless, Pink Tooth, Dwarfism, and Imperfect Skin. Red hair color was deleted from the list in 1969 when red & white animals became eligible for registration.

The leadership role of the Holstein Association in publicly identifying the carriers of undesirable recessives has become a major action of the era in upholding the genetic health of the breed.

Robert Kessen managed the Association's finances for 30 years.

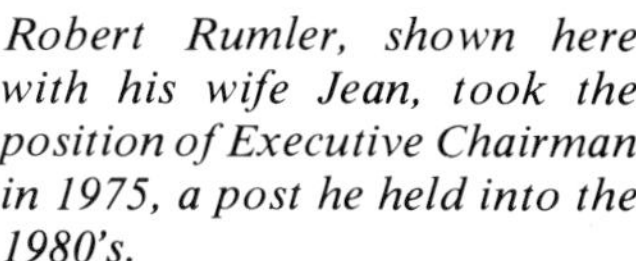

Robert Rumler, shown here with his wife Jean, took the position of Executive Chairman in 1975, a post he held into the 1980's.

Charles Larsen became Executive Secretary in 1975.

Zane Akins, the seventh Executive Secretary of the Holstein Association.

While expanding services at home during the 1970's, the Holstein Association spread the name "U.S. Holstein" across the World.

Holstein Association's 1985 centennial year officers and directors.

Accomplishments And Changes

Many production records were broken during the seventies as the Holstein breed charged ahead. In 1971 a new world's record for milk production was set by Skagvale Graceful Hattie, bred and owned by Tenneson Brothers, Sedro Woolley, WA, when she completed a 365 day 2X record of 44,019 pounds milk and 1505 lbs. butterfat.

This record was broken in 1974 by Breezewood Patsy Bar Pontiac, bred and owned by Gelbke Brothers, Vienna, OH (45,770M 2191F), becoming the first 2000 pound butterfat record in the United States. But that milk record was broken late in the year by Mowry Prince Corinne, bred and owned by Clarence and Ken Mowry, Roaring Spring, PA, with 50,759 pounds of milk —the first cow ever to produce 50,000 pounds in a year.

The next year the world's record for milk production was set, a record that stands today. A six-year-old registered Holstein, Beecher Arlinda Ellen, owned by Harold Beecher, Rochester, IN, produced 55,661 pounds of milk and 1572 pounds of butterfat. She epitomized the genetic advances that had been made by the breed.

The Association said farewell to several longtime employees during the 1970's. Robert Stebbins, who had served the organization for 46 years, many as Superintendent of the Registry Department, retired in 1973. Florence Dawson, who had more than 48 years service, retired as Chief Supervisor of the Registry Department the following year. And Robert B. Kessen, a key manager of the Association's financial affairs for 30 years, retired in 1980.

Midway through the decade, the Office of Executive Chairman was created to strengthen the approach to policy making, and to solidify the position of the Association in the national and international agricultural community. Robert Rumler took on this position which he held into the 1980's.

Charles Larson, who had served the Association since 1955 in many capacities, including 10 years as Assistant Executive Secretary as well as Executive Vice President of HFS, Inc., assumed the position of Executive Secretary in 1975.

Zane Akins, who had begun his Association career as a program director in Wisconsin and then successfully launched the Holstein Sire Development Service, and who had been Administrative Assistant for a little more than a year, took over the helm of the Association in 1978. Akins, only the seventh person to be Secretary, was 37 when he was appointed to lead the worldwide organization.

The 1970's were an exciting expansion period for the Holstein Association. The organization had launched a host of new service programs, had implemented new genetic evaluation methods, and had continued its steady growth in a period when dairy cattle numbers were continuing to decline. It had weathered inflationary financial storms by raising fees when it had to. And while expanding services at home, the Holstein Association had spread the name "U.S. Holstein" across the world.

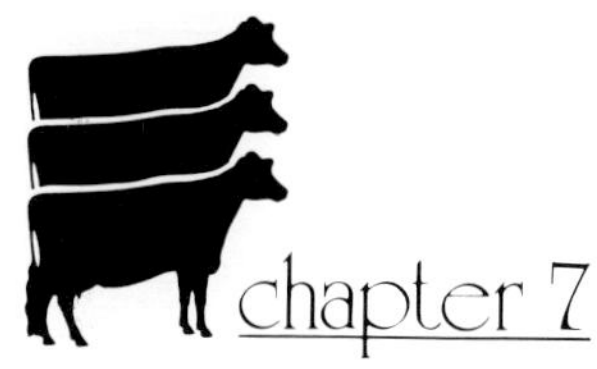

chapter 7

A Global Breed Organization

Each leader of the Holstein Association has brought new ideas and his own management style to the organization. F. L. Houghton was a "one-man band" for many years, and with his forceful leadership style, shaped the organization in the early growth years. H. W. Norton preferred to use the committees of the Board of Directors in an active management role. Robert Rumler operated, with great success, using the Chief Executive Officer technique. Zane Akins brought a participatory style of management to the Association when he took over as Executive Secretary in 1978. He established a new organizational structure which called for division heads to assume major shares of responsibility, and began a number of management initiatives to lead the Association through the rest of its first century and beyond.

Management Initiatives For The 1980's

Robert Rumler had brought long-range planning to the Holstein Association, preparing a series of three 10-year plans that mapped the organization's course through the postwar growth years. Executive Secretary Akins implemented a more formal approach to planning, shortening the interval to five years, and using a middle management planning team. Each July the Board President, Vice President, and Executive Secretary sat down with the division directors and, together, identified major areas where new results should be achieved. Then each division prepared an action plan to achieve specific results. This "management-by-objectives" approach to management has been used since 1979.

The Association also started to become more aggressive in marketing its services. Beginning with market research and studies on the attitudes and

Executive Secretary Zane Akins, with the guidance of William Van Dusen, agricultural consultant (right), implemented a series of five year management challenges.

Public Relations Director Ben Coplan used new techniques to promote U.S. Holsteins.

Direct marketing began in 1980 when trained staff began fielding questions from breeders and selling Association services by telephone and mail.

opinions of members, the public relations department, under the leadership of Ben L. Coplan, supplemented the more traditional methods of the past (advertising in the trade journals, press releases to newspapers) with modern public information and marketing techniques. Professionally-designed brochures described the service programs of the Association while at home and abroad, glossy posters of sleek Holsteins extolled the merits of the breed. "This is a unique marketing opportunity," Coplan explained. "We're selling the future of the breed."

Rather than waiting for dairymen to decide to use the services, the Association began direct marketing, using both mail and telephone. And the new techniques worked. During the last eight years of the Association's first century, revenue tripled, as did participation in nearly all the herd improvement programs. The new services and products that had been introduced in the 1970's allowed the Association to get away from relying on registration and transfer fees for its financial lifeblood, as it had for nearly 90 years — now it derived nearly half its revenue from other services.

The computer capacity of the Holstein Association was upgraded throughout the early 1980's. The first on-line system for customer records was installed in August 1980. By July 1982, registry records were on-line and 70 computer access terminals were in use throughout the Association offices in Brattleboro. When a new calf's records were entered, the computer system performed 128 edits, verifications, and cross checks in less than two seconds, and if the information was compatible, accepted the animal into the electronic herd book. In September 1983, a one-day processing cycle of registry, inquiry, and pedigree data was implemented departing from the previous weekly run. The speed and capacity of the computer in 1983 was five times that in 1980, as the Association kept up with emerging technology. The first Linear Sire Summary book was processed in early 1984 while in June of that year, the computer was linked directly to the service blood-typing lab in Texas.

In 1980 the Association field staff, which had doubled in size in 10 years, had grown to the point where it was becoming impossible to manage from a central office. The widely scattered field staff needed administrative support near them,

Guide for Permanent Holstein Identification
HSDS
Holstein Sire Development Service
YOURS FREE!
When You Mail This Card
YES, PLEASE HAVE A CONSULTANT CONTACT ME
FIND OUT HOW THE HOLSTEIN ASSOCIATION CAN HELP YOU BUILD A BETTER BUSINESS
Your Grade Holsteins Over 24 Months of Age May Still Be Eligible For Permanent Identification
Experience the exciting story of
"THE WORLD OF THE U.S. HOLSTEIN"
Give This Card To a Neighbor
PRINTS OF FILM ARE AVAILABLE FOR PURCHASE

The Association office is a study in contrasts. Here, a massive painting of the True-Type bull overlooks registry staff and their modern CRTs.

to allow them to concentrate on their service duties. Restructuring of the field force was the answer. Five Area Managers were selected from the field staff of 50 classifiers and consultants, and Area Offices were established in Fresno, CA; Altoona, WI; Shelbyville, KY; Kansas City, MO; and Harrisburg, PA. This regional delivery system for on-farm service to dairymen expanded the full-service concept begun in the 1970's, vastly improving the quality of fieldwork. Better management, better support, and keeping the field staff closer to the membership all contributed to the effectiveness of the Area Office program.

The original Area Managers. From the left: Neil Hammerschmidt, Mel Hertzler, Ron Shaver, Bob Cain, Bill Lupo and Jim Pound, Director of Field Services.

Holstein Association
1985 Organizational Structure

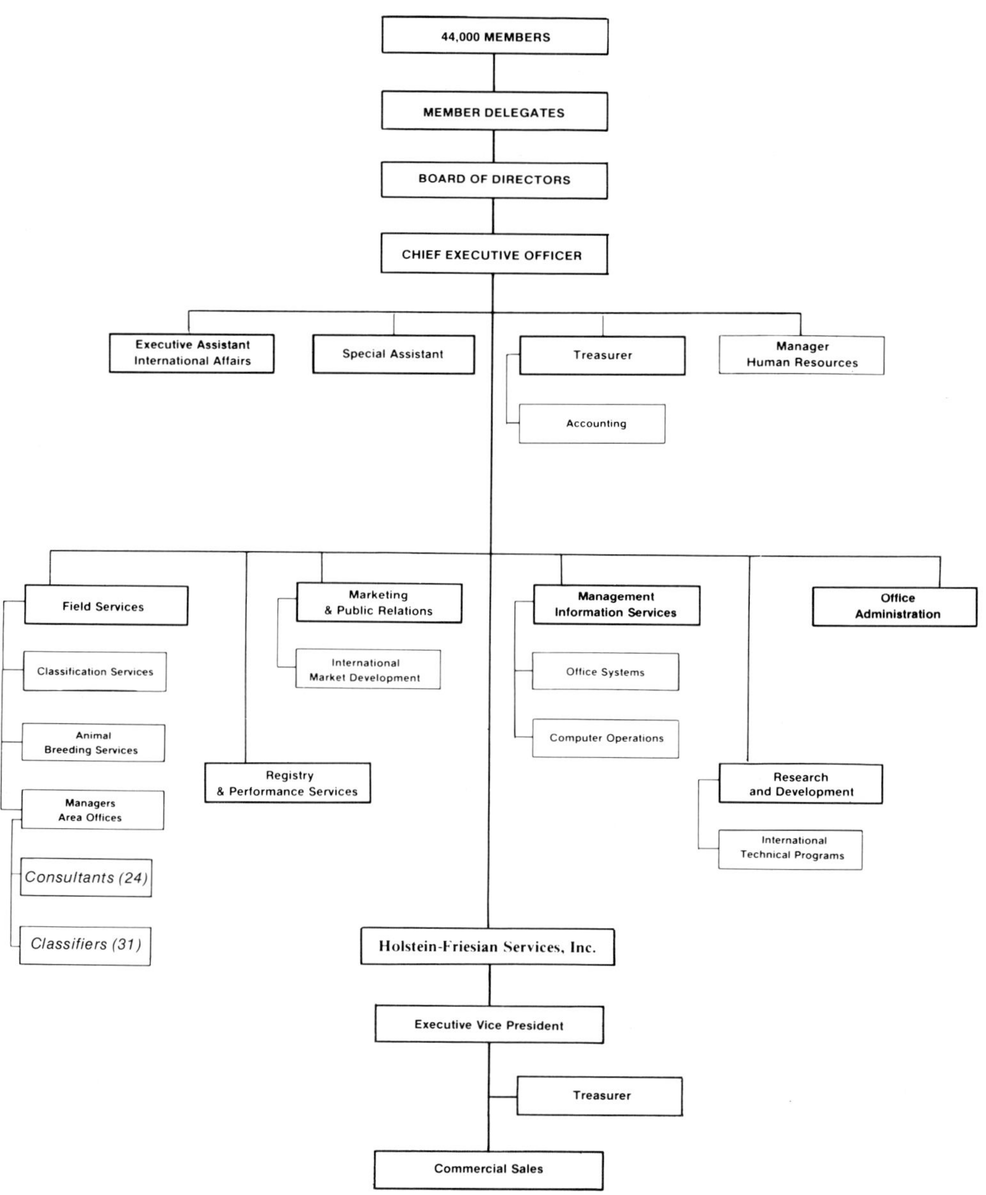

Holstein-Friesian Association of America
Brattleboro, Vermont 05301

A Holstein Association classifier demonstrates the technique of Linear Classification to a group of foreign classifiers.

Linear Classification

One of the first new programs the Area Offices administered was a new classification system, Linear Classification. Since the introduction of Descriptive Classification by the Holstein Association in 1967, more than four million registered Holsteins had been classified in the United States and thousands more overseas. But breeders were using a program, Descriptive Classification, which had been designed to be a mating tool, and now needed more precision to make breeding decisions in AI programs. With advanced computer capability and better genetic evaluation procedures now available, it was time to revamp the classification system.

In 1979 the Holstein Association and the National Association of Animal Breeders (NAAB) worked jointly on a research project to develop a single type evaluation program for U.S. cattle. Its purpose was to more accurately measure specific type traits, getting away from the concept of "what is ideal?" thinking. The research project considered the best features of all current programs and then field tested both the descriptive procedures and a new "linear" system. Over a 15-month period, Holstein Association classifiers evaluated over 20,000 cows in 490 herds using both systems. In 1981, the Board of Directors adopted the Linear Classification System, to become effective January 1, 1983.

The new classification system abandoned the comparison with the "ideal" and went, instead, to a system which measured the degree or magnitude of a given trait. The system was designed to serve two primary functions:

1. Genetic Evaluation — to measure the magnitude of economically important type traits.
2. Merchandising — to assign a final type score that is easily understood and respected throughout the industry, for recognition and marketing purposes.

Linear traits were divided into 15 primary and 14 secondary traits. The primary traits, which were felt to have economic value, were divided into five categories: Form, Rump, Legs and Feet, Udder, and Teats. Secondary traits, used at first for research, were recorded only at extreme values. Linear Classification, while retaining the time-tested concept of final score, provided the breeder with a linear rating of each functional trait on a 50-point scale of biological extreme.

Embryo Transfers

In March 1974 the first Holstein eligible for registration that had been born as the result of an embryo transfer (ET) was registered at the Holstein Association's office in Brattleboro. ET, which would grow rapidly in the late 1970's, posed some of the same documentation and credibility problems that artificial insemination had 30 years before. The new technology also brought a new type of owner, the outside investor, into the registered Holstein business because of the tax advantages.

Embryo transfer had been performed on laboratory and farm animals on a limited basis since the 1890's but had only become commercially feasible in the mid-1970's. As improved techniques developed, the number of ET calves increased, but nowhere near the same rate as AI calves had during the early growth years of the AI program because, unlike AI, the procedure was not cost-effective for most commercial dairymen or for many owners of registered Holsteins. However, there were several distinct advantages to ET:

1. More progeny could be obtained from selected females.
2. Calves could be obtained from cows too old to sustain a pregnancy.
3. Holstein genetics could be shipped abroad in nitrogen tanks instead of jets with less chance of disease problems.

The basic disadvantage to ET for most breeders was the cost of the process. The shots to induce superovulation, the specialized skill needed to harvest embryos, and the uncertainty of the number of viable embryos made the process expensive and high risk. Yet, for some it was a profitable way to market Holstein genetics. And the registration of ET animals began to climb:

ET Registrations

	Females	Males
1975	10	5
1976	92	41
1977	161	100
1978	321	295
1979	774	705
1980	1957	1598
*1981	3333	2424
*1982	4854	3155
*1983	4059	2742

*These figures will change. Animals from these years may still be registered at a time in the future.

The advent of ET opened up marketing opportunities for the Holstein Association, using their business corporation, HFS. The international embryo transfer program began for HFS in 1979, when a five-year technical assistance agreement with a leading breeder in Spain was signed. In late November, two fancy Holstein cows ("Preen" and "Ginger") were shipped to a farm south of Madrid. The cows, both granddaughters of the sire Osborndale Ivanhoe, were exported for the specific purpose of embryo transfer.

Noting the turning point in the export of dairy genetics, Association Executive Secretary Zane Akins said of the HFS shipment, "ET will become even more significant as domestic cattle prices rise, transportation costs skyrocket, and animal health requirements become more difficult to fulfill." HFS pioneered the exporting of fresh embryos in 1981 and began the domestic marketing of embryos the following year.

ET also presented new registration and credibility challenges for the Holstein Association. In order to maintain close control over the lineage records when dealing with fresh, frozen, or even divided embryos, new procedures had to be developed. The donor dam had to be blood typed as well as every ET calf, and special requirements were set up to label all frozen embryo containers. Once again, like the early days of AI, the blood-typing program of the Association, which had been nurtured so well by Richard Nelson and his staff, proved its value.

World Friesian Conferences

Breed officials from nine countries attended the Diamond Jubilee celebration of the Holstein Association in Syracuse in 1960. The day after the convention, they met to discuss mutual problems and interests. They decided to hold an international conference every four years, and so the World Friesian Conferences were begun.

The Dutch, proud of the homeland heritage of the breed, sponsored the first conference in Amsterdam in 1964. In 1968 the Holstein Associations of Canada and the United States were co-hosts of the 2nd conference in Harrisburg, PA. Subsequent conferences were in Italy in 1972, Great Britain in 1976, Germany in 1980, and Mexico in 1984. In 1984, 20 of the 38 Holstein or Friesian herd book organizations from throughout the world participated.

By design, there has been no formal structure to the organization — the host herd book society provides the acting chairman for the conference. As international operations have become more commonplace, there is, in 1985, a movement to make the World Friesian Conference a more structured group with an elected leader, which would meet more frequently than in the past.

International Ethics

As U.S. Holsteins spread across the globe, concerns surfaced regarding the integrity of pedigree records. As millions of dollars worth of Holstein genetics moved throughout the world each year, some of the same ethical problems that the Association had encountered domestically were encountered in international trade.

Holstein-Friesian Services, Inc. had been established in the late 1960's to help

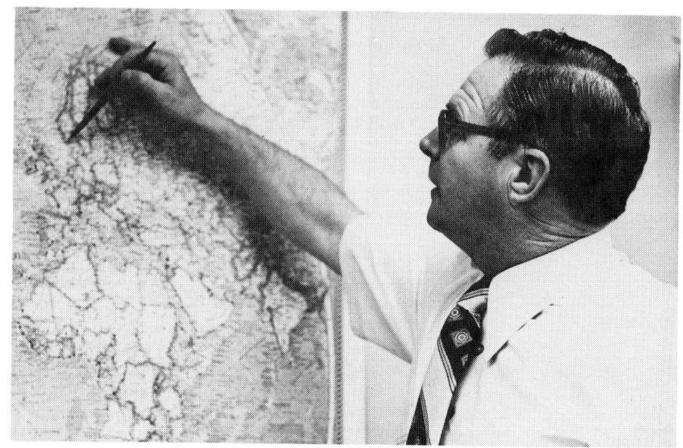

Maurice Mix, International Affairs Assistant, has been instrumental in arranging International Operating Agreements with 13 foreign herd books.

bring a standard of quality control and ethics to the export scene. Yet, as more and more exporters became involved, the incidences of fraud increased. The problems were highlighted by a case in 1983 in which 207 head of grade Holsteins were shipped to South Korea with false registration papers. The forged Holstein Association registration certificates and pedigrees which accompanied the animals were well prepared. The counterfeiters, who were taken to court by the Association, were discovered only because of a technical error in registration numbers. The case pointed out the sophistication of potential fraud in international commerce, and the need to set up controls to maintain the integrity of the Association's registration certificate abroad.

The Association had first tackled the whole international ethics problem with International Operating Agreements — a herd book to herd book system. The purpose of the agreements was to assure any sister herd book that accepts U.S. Holsteins for entry that the documentation on lineage, previous and present ownership, and performance is accurate and that the papers accompanying the Holstein genetics or animals are genuine.

The first such International Operating Agreement was signed with Italy in 1981. Through 1984, 13 sister herd books have signed agreements. They are: Australia, Chile, Colombia, Costa Rica, Ecuador, Italy, Japan, Kenya, Korea, Mexico, Peru, Spain, and Venezuela.

Success In Hungary — An Example And A Model

While many foreign countries have benefited greatly from the Holstein Association's technical assistance program and U.S. Holstein genetics, the results achieved in Hungary during the 1970's and 1980's were unmatched anywhere else in the world. "No other country in the world has increased milk production more rapidly than Hungary," reported a Hungarian official in 1983 at the first International Holstein Conference in Budapest, where the impressive results of a decade of technical assistance were announced:

> *Per capita consumption of milk products in Hungary nearly doubled, yet the country became self-sufficient in meeting its dairy needs.*

Hungarian medallion honored the new Holstein milk production average of 11,243 lbs. (5000 liters) per cow.

Zane Akins presents a commemorative plaque to Hungarian Minister of Agriculture and Food Jenos Vansca celebrating the success of the 10-year dairy cooperation agreement between the two countries — August 1983.

Dr. Ray Kliewer with the first Chinese visitors to the Association.

More milk was produced from fewer cows in large modernized dairy operations — average production per cow jumped 83% — the national herd was cut 23% — feed, labor, and overhead costs were reduced.

Holstein-Friesian Services, Inc. shipped the cattle and sponsored technical cooperation to do the job. In 1973, HFS shipped their largest number of animals to Hungary which bought 5128 Holsteins and Herefords. These exports began a linkage between the Holstein Association and Hungary which was to flourish for the next decade. It was the start of a nationwide dairy herd improvement program in Hungary, primarily done through HFS, that achieved impressive results.

In the early seventies the Hungarian Ministry of Agriculture and Food had decided to change the country's breeding program from a traditional dual-purpose system to one of better livestock production and improved milk production. Justification for this approach was strongly supported by a feasibility study conducted jointly by the Holstein Association, the Hereford Association and USDA. From this background, the Holstein Association presented a comprehensive plan to Hungary which was accepted.

Holsteins from the United States were chosen to be crossbred with the native Fleckvieh cows. A technical assistance program was begun between Holstein-Friesian Services and Hungary's two agriculture branches: The National Center of State Farms and The National Feeding and Animal Breeding Inspectorate. During the 1973-1983 period, HFS shipped 18,000 U.S. Holstein heifers and 150 bulls to Hungary. Over a million units of frozen semen and embryos were also involved.

The technical assistance needed to get the job done was carried on under the leadership of Dr. Ray Kliewer, Director of Research & Development for the Holstein Association. U.S. dairy specialists worked with Hungarian specialists and farmers on all aspects of herd improvement. Management was a key to success. Kliewer made 34 trips to Hungary over the 10-year period. In all, 22 specialists from the United States spent a total of 551 days in the country providing technical aid.

As has happened in other countries, as the Holstein breed grew it spawned a number of support industries. In Hungary a whole agribusiness system developed around U.S. Holsteins. The expanded dairy industry supported companies that provide seed, fertilizers, chemicals, farm and dairy processing equipment, feeds and premixes, and storage and handling equipment. The Hungary success story accentuates the impact of the Holstein Association on the worldwide dairy industry in the 1980's.

A Century Of Progress

The Holstein Association has grown from a small band of breeders into an international force in agriculture. In a century of progress, the U.S. Holstein breed has become the premier dairy cow of the world, expanding from several hundred dual-purpose animals to over 14 million registered animals. The growth has been phenomenal since World War II: It took 40 years, 1885 to 1925, to register the first million animals; 14 years to register the 2nd million; 9 years to register the 3rd million; 7 years to register the 4th million; 5 years to register the 5th million; and now, a million animals are registered in just over two years. And all this took place in an environment which saw the number of dairy cows in the country diminish sharply.

The number of Holstein registrations per cows being milked has risen constantly in the last three decades so that, in 1984, there is one registered Holstein for every 21 Holstein cows in milk, truly a remarkable achievement.

Over its history, the Holstein Association has helped to establish and maintain a steady financial base for its members, assuring them and the industry of the stability of the organization. By adopting a "pay-as-you-go" policy, it purchased the Brattleboro building and built both the 1950 and 1961 additions without borrowing funds. The Reserve Fund, set up in 1902 to cover the lean years, helped the Association weather the Depression years. Since then, it has been built up and maintained as a financial anchor, untouched in the day-to-day affairs of the organization, bringing the element of fiscal solidness that has added credibility to the lineage records of the breed. Breeders in the United States and throughout the world have known that they would have a sound, ongoing herd book organi-

Holstein Consultant Richard Howe directed a special project, which began in 1984, to upgrade the herd of the Brattleboro Retreat. The herd will be used to demonstrate the benefits of using the Association's services and expertise.

zation to work with, regardless of the economic conditions of the nation, or the world.

The history of the Holstein Association is a roll call of dedicated people who, each in a unique way, contributed to the advancement of the breed. We have covered the contributions of the visionary leaders of the Association such as Hoxie, Houghton, Norton, Rumler, and Akins, who could see the Association as much more than a record-keeping society, as a dynamic leader of the dairy industry. But the dreams of these leaders could not have been realized without the painstaking work of thousands of others.

Generations of file clerks, keypunch operators, and computer terminal operators have formed the backbone of the Holstein Association since the start. Whether the task involved carefully drawing the color markings of a heifer on a registration certificate, the entering of registry data into the computer, or handling a call from a dairyman in Oshkosh, the care and accuracy of these employees have done much to bring the Holstein Association to the vaunted position of respectability it enjoys today.

Some of the first Holstein cattle traveled to foreign lands with missionaries. The Holstein Association has had its own group of missionaries who, since the Depression, have traveled the rural roads of the United States spreading the word about registered Holsteins. Fieldman Allen Crissey, whose influence spread from the northern reaches of Maine to the Carolinas was one of many. Another was the "Dean of Field Service," Robert Cain, who began his Association career in 1952 by loading a pickup truck with a Holstein display at the Brattleboro office and driving straight for the fair in Ionia, MI. Cain did this for three years, hitting the major state fairs and then returning to his Deep South field area (northern Kentucky to the Florida Keys).

Yet, the real strength of the Holstein Association has come from its membership, now including youth members, over 53,000 strong. From the new

Maurice Mix discusses U.S. Holsteins with the president of the Korea Animal Improvement Association and a Korean classifier.

junior member with her first 4-H calf to the seasoned veteran of years of annual convention debates, members have provided the ideas and the energy to drive the Association to greater challenges and accomplishments.

And the accomplishments have been extraordinary. The Association spearheaded the spread of U.S. Holsteins across the country and the globe by helping to develop a genetically advanced breed that is valued for its production and longevity at home and abroad. As it entered the 1980's, the Association, in a written goal, charted its course as it begins its next century of operations:

> "Our mission is to provide genetic, management and marketing services of the highest quality to Holstein dairymen while maintaining the credibility of information, the integrity of the breed, and while also developing a business and political climate for the dairy industry that will provide optimum economic opportunities."

It has been a century of challenge and change. The Holstein Association responded to the challenges by helping to develop a product, the U.S. Holstein cow, that is recognized as the premier dairy animal in the United States and the world. In doing so, as Robert Rumler said in the Foreword, it helped to make the transition in the breeding of dairy cattle "from an art to a science."

Korea has been a major importer of registered U.S. Holsteins.

"Our mission is to provide genetic, management and marketing services of the highest quality to Holstein dairymen while maintaining the credibility of information, the integrity of the breed, and while also developing a business and political climate for the dairy industry that will provide optimum economic opportunities." **Holstein Association**

Over the last century, U.S. Holsteins have spread from New England

to California

from North to South

Maggie Murphy

and from East to West.

Danny Weaver

Holstein Association staff have helped lead the way,

promoting the breed across the globe,

from Hungary and Yugoslavia,

to China,

and Japan.

Holstein Association members, 44,000 strong,

have had visionary leaders

who have protected the integrity of the lineage records stored at the Brattleboro, Vermont, home office.

While public sales
have provided excitement
for many,

and the show ring has brought thrills and reward,

there are many intangible benefits

for the breeders of U.S. Holsteins.

PART 2

Holstein Activities

Introduction

While the Holstein Association was instrumental in leading the way during the last century, many others also helped bring the U.S. Holstein breed to its premier position in the world. The activities covered in this section are some of the things that make breeding Holsteins rewarding, both to a breeder's satisfaction as well as his pocketbook. Many are traditions as old as the breed itself and this patina of history adds an element of continuity in the "hi-tech" world of the 1980's. For a breeder interested in building cow families, of leaving a genetic legacy, it may be pleasing to compare blue ribbons won by his animals with those of his father, or grandfather. In the auction ring, a nod of the head may still buy a $5000 cow, the waggle of a leg raise a bid — movements that worked just as well a half century ago. Yet, while many of the activities of the registered Holstein industry have deep roots, they are as current today as ever. Holstein breeders are in the business of selling genetics as well as milk, and these activities help get the job done. The sale ring sets the prices for public and private sales, the hundreds of Holstein shows present unique merchandising opportunities, while the breed publications are a breeder's "Yellow Pages." They all have an integral part in the history of the Holstein breed.

chapter 8

Public Sales

The excitement of a registered Holstein auction is catching. It can infect a 4-H youngster watching her homegrown heifer being sold, a savvy breeder dispersing a herd carefully built up over the years, or a veteran cowman bidding for absent buyers. Each is drawn to the ebb and flow of action of the sale ring. And that excitement is the magnet that has held some breeders in the registered Holstein ranks. One veteran breeder put it like this: "With grade cows, all you can do is get up each morning and milk them. With purebred Holsteins, you classify them, test them, show them, and sell them. That's where the fun is."

The sale ring has been a barometer of the Holstein industry for the last century, as sale prices have risen and fallen with the overall interest in the breed. Until recently, it was a cyclic barometer, following the supply and demand of Holsteins

Editor's Note: No adjustments have been made for sale prices to common base so a strict year-to-year comparison may be erroneous. The reader is cautioned that all figures are quoted in dollars for that year.

Barney Kelley

Bob Haeger

Early Auctioneers

Col. D. L. Perry

Walter Andrews

by farmer/breeders across the nation. Public auctions have been stirring theatrical productions at times, featuring euphoric demonstrations in the early 1900's and lavish productions in exotic settings in the 1980's. What color the auction has brought to the industry! Auctioneers whose names are part of the lore of the sale ring include Barney Kelley, Bob Haeger and Gene Mack of the early days; the persuasive Col. D. L. Perry of Ohio; the silver-tongued Walter Andrews; and the popular C. B. Smith.

Since World War II, public auction sales have reflected the dramatic story of the spread of Holsteins across the country and the globe. Public sale results have followed the advances in production testing and sire selection with the highest prices being commanded by those genetically superior animals, and their progeny, which have demonstrated the ability to transmit their best traits. While show winnings and classification scores are still important to many, in the last two decades the sale ring has rewarded those breeders who used modern tools to make selections.

Yet, in spite of being financially sensitive to genetic developments in the breed, the public auction is a paradox. In an era of electronic fund transfers, sophisticated tax shelters, and embryo transfers, the traditional public auction has remained virtually unchanged — one of the few aspects of the Holstein industry to do so. Auctioneers still look for a head nod or foot waggle to signify a bid, and newcomers sit still in fear of inadvertently bidding on the animal in the ring by scratching an ear at the wrong moment. "Pedigree men" extol the virtues of the heritage of the animal being auctioned and tell of the genetic advances that future matings can achieve. But the animals being sold have changed dramatically, evolving from the first beefy imports to the sleek angular high producers that now command top prices. And the financial arrangements, especially at the major sales, have changed as the stakes have risen — now many sellers of top animals virtually have the sales arranged before the animal hits the ring.

The auction is still the place to go for the farmer who wants to escape the loneliness and drudgery of the dairy farm, a place where one can participate by consigning, bidding, buying, or just watching the action. And there is constant action. Whether in a posh Reno casino setting or at a dusty midwest county fair, the public auction still stirs the pulses of buyer and seller as it has for over 100 years. The tension in the crowd, the veteran auctioneer scanning for bids, the litany of the buying and selling — all of it could be a page out of the first Brentwood sale in 1920. The public auction, always the glamour event of the industry, is still "where the action is."

But, while it is stimulating and enjoyable to watch bidders match wits with the skilled auctioneer on the stand, it is also important to recognize the role that the sale ring has played in bringing U. S. Holstein cattle to the position of prominence that they now enjoy — by providing Holstein owners a ready source of cash, by helping to establish fair prices, and perhaps most of all, by helping to popularize the breed.

The auction sale can provide a Holstein breeder with a ready means of converting some of his registered animals into cash. For the dairyman whose financial worth consists of his farm and herd, this is important. While it may take some time to dispose of real estate, the registered Holstein herd has proven to be the same as (and, during inflationary times, better than) money in the bank. For reasons besides milk, a purebred Holstein can truly be considered a "liquid asset."

A second benefit that public auctions have provided to the Holstein industry is that of establishing the current market value of cattle. Since seven out of 10 registered Holsteins sold each year are the result of private sales, auction prices can be particularly helpful in guiding prices at these private sales. A breeder who attends a few auctions in his area can compare the animals being sold with his own, and know within reason what his own animals are worth.

And most importantly, auction sales of U. S. Holsteins have been instrumental in promoting the breed. Good news travels fast. Over the years, the pedigrees, the milk production records, and the sale prices of registered Holsteins have influenced many a dairyman to switch from the colored breeds. They also have influenced some commercial dairymen to join the ranks of registered Holstein ownership.

A promotional sale affords dairymen the opportunity to market selected animals from their herd.

Promotional Sales

Promotional auction sales have played an important role in marketing registered Holsteins for better than a century. In the late 1800's, some importers used this method to find buyers for their Dutch cattle. An auction sale staged in New York on March 17, 1885, by Peter C. Kellogg & Co., turned out to have a major impact on the Holstein breed. The reports from that sale vividly point out the reasons the Dutch Friesian Association and the Holstein Breeders Association joined forces (Chapter 1). First, one of the consignors was Sluiter Brothers of Groningen, Holland, who had angered U.S. breeders by infringing on American trade. And secondly, there were complaints that the quality of some of the consigned animals was poor. These issues hastened the formation of the Holstein-Friesian Association later that spring.

When, in the early 1900's, sale management became a profession in the U.S., managers began conducting "Breeder's Consignment Sales" which are now called promotional sales. Starting in 1903, consignment sales were held in conjunction with the annual meetings of the Holstein-Friesian Association. In 1916 these were designated "National" sales and received the endorsement of the Association.

Another important series of sales also started that year. The Allegany-Steuben Holstein Club in New York state held the first county or area Holstein club sale. Believed to be the oldest club sale series in the country, it is now in its 70th consecutive year. Since then, many county Holstein clubs have held their own sales or joined together in district sales. State Holstein Associations sponsor promotional sales to help merchandise the cattle of their members. This "grass-roots" series of public auctions has been a major promotional effort for the U.S. Holstein breed.

There are many benefits to a promotional sale. It is an excellent merchandising tool for individual breeders and sale managers, allowing dairymen the opportunity to market selected animals from their herds. And history has proven that the consignor of a choice Holstein often receives recognition for the animal and the herd worth far more than the sale price. But the benefits are not just limited to the seller. The prospective buyer has an array of prescreened animals to choose from, carefully selected for production, pedigree, and type. Without having to spend extensive time and travel to search out animals of this caliber, the buyer can, after studying the advance advertising, determine the good ones to bid on and then either attend the sale in person or authorize an agent to make the purchases.

Dispersal Sales

Nowhere else do the dynamics of the public sale come more into play than at a dispersal sale. To the breeder/dairyman whose business life has been dedicated to the farm and herd, a dispersal (a complete sale at auction) of a herd is one of the most important events in a career. A culmination of a lifetime's effort, the sale may well determine the breeder's financial future, whether retirement or in some other endeavor. It does not matter whether the herd is of national or statewide prominence: in the eyes of the breeder, this dispersal is the most important one ever held. There is a lot riding on the gavel of the auctioneer.

On the buyer's side, there is the potential of reaping the benefits of that breeder's handiwork, of taking promising or proven animals that have been developed in the breeding program, and bringing them up to new genetic heights. And while dispersal sales sometimes have the poignancy of the "end of an era" about them, they also have the excitement of a genetic launching pad for further herd improvement as the animals are dispersed into new herds.

In more recent years, dispersal sales have been held for many other reasons. Sometimes the sale is a result of the dissolving of a partnership or to settle an estate; at other times it is strictly a business venture, where an owner has built up a herd through purchases over a period of time with hopes of "cashing in" on the investment, counting on cow knowledge and marketing ability to turn a profit. And in the last 30 years, partial dispersals, "herd sales," have become more frequent. Whether to reduce the size of a herd, to create a cash flow, or just to take advantage of a favorable market, it is simply another way to market registered Holsteins.

There are many public sales each year that are unreported, but in many respects, they are equally important as the attention-drawing top sales. These local sales move most of the registered animals sold at auction each year, and, since many handle Identified Holstein females and other grade Holsteins, they are sometimes the place where the commercial dairyman purchases his first registered animals. These sales provide a fine "cash crop" opportunity for the local farmer/breeder and, if run by a Holstein club, often serve as an important fund raiser for that group. Local public auctions are true market tests and feature cash purchases.

While it is satisfying to sell an animal that you have bred and raised at the top of a local sale, it is likewise satisfying for top breeders to achieve success in national sales. There are few finer ways to advertise a herd or a breeding program than to place a top animal in a well-promoted public sale. A high quality animal reflects credit on all concerned: the breeder, consignor, buyer, and the sale manager. Whether in the ring at the Steuben County Fair in Bath, NY, or in Madison Square Garden, public auction sales of Holsteins have done much to spread the word about the breed. They are, in no small part, a major factor in the incredible history of the U.S. Holstein.

Early Sales Of Holsteins

Public sales of Holsteins began over a century ago. Many of the prominent foundation animals of the breed and their progeny were sold at auction. Men like Thomas B. Wales, the first Secretary of the Holstein-Friesian Association, were instrumental in popularizing the breed through the sale ring. While there are reports that several public auctions were held as early as 1879, the first landmark sale on record is the 1883 sale at the International in Chicago. Wales, who was secretary of the Holstein Breeders Association at the time, had created quite a publicity splash by moving his herd from Massachusetts to Iowa the previous year. Now, he drew further attention to the breed by drafting 37 animals for the sale, which drew buyers from Ohio, Illinois, Kentucky, Wisconsin, Pennsylvania, and Iowa. The outstanding offering was the yearling heifer, Mercedes 3d, daughter of Mercedes, purchased by Smiths, Powell and Lamb for $4200, an astronomical price at that time.

Five years later, Wales, who by now was heading the Holstein-Friesian

Sale day is exciting for buyer and seller alike.

Association, dispersed his herd in the major Holstein auction of the 1800's. Secretary Wales, who was soon to run into trouble through his heavy-handed operation of the Association, was instrumental in popularizing the breed during a difficult period. The glut of poor imports had hurt the fledgling industry, and the new breed association was fighting to gain a toehold. The Wales herd, with a nationwide reputation, was shipped from Iowa to Cleveland for the sale, and attracted buyers from nine states and Canada. The renowned butter test winner who had been imported by Wales, Tritomia, as well as several of her sons and daughters, found new homes. Sixty-six head, selling for an average of $455, spread the reputation of Holsteins and some famous genetic lines to new parts of the U.S.

This highly successful sale stimulated several more the following year. The *Holstein-Friesian Register* reported eight sales in which 255 head averaged $116. The outstanding sale in 1890 was the dispersal of the W. B. Barney & Associates Home Farm in Iowa where Colantha, who would go on to be one of the famous cows of the breed, was sold to Gillett & Son of Wisconsin for $540.

For the balance of the 19th century, the tide of Holstein fortunes fell as economic conditions and stiff competition from the other breeds dampened enthusiasm for the breed. This lack of interest was reflected in the few auction sales that were held — prices were often below $100 for a good registered Holstein cow. Sales for all of 1900 show an average of $60 per head. But the advent of Advanced Registry (AR) testing would turn Holstein prices upward as emphasis started to shift from how cattle looked to how they milked.

Higher Production — Higher Prices

Interest in Holsteins and sales prices picked up as the new century began. AR testing was gaining popularity and when Lilith Pauline De Kol set a new world's record for the seven-day butter test (28.23 lbs.) at the 1901 Pan American Exposition in Buffalo, her owner was offered, and refused, $2000 for her. The

S. D. W. Cleveland was the breed's first professional sale manager.

first Breeder's Consignment Sale, put on by W. C. Hunt the following year, marked the turning point in the depressed market that had prevailed so long. Here was the first time that breeders were willing to put a price on performance and pay substantially more for AR animals. 1903 saw the trend upward continue with prices reported as "the highest since the flush times of the early eighties." Sadie Vale Concordia became the first cow to reach the 30-lb. official butter test record that spring, and when the Brothertown herd was dispersed later that year, the 63 head brought an average of $280, the highest in years. Following the annual meeting of the Holstein-Friesian Association in Syracuse, Averill and Gregory dispersed their famous 159 head herd at high prices. The famous "28-lb." cow, Segis Inka, topped the sale at $1600.

The following year, a consignment sale was notable for two highlights. It marked the start of the first professional sale manager, S. D. W. Cleveland of Syracuse, who put together a sale featuring cattle from most of the leading eastern breeders. It also marked the last sale of imported Dutch cattle, which had fallen into disfavor because of poor production records. These particular imports, however, had been personally selected from Holland by Ward Stevens and T. A. Mitchell and drew good prices, about $100 over the sale average. But they would be the last Holsteins brought from Europe — an embargo was placed on such imports the following year due to foot and mouth disease. As noted in Chapter 1, there have been no European importations since.

Sale activity continued to increase as interest in Holsteins sharpened. The first Holstein to break the five-figure barrier was King Segis Pontiac Alcartra, who sold as a young calf for $10,000 in the Breeder's Consignment Sale of 1911. Two years later, Holstein public auction sales surpassed the million-dollar mark for the first time (4737 head for $1,178,816).

But the year 1913 was important for another reason — it marked the first sale of the breed where every animal was offered under a strict 60-day guarantee against tuberculosis. The guarantee, which would become standard practice, marked an end to the "caveat emptor" approach to T.B. that faced buyers of Holsteins. Prior to 1916, tuberculin testing was on a hit-or-miss basis. The first Holstein auction which advertised all animals negative to the T.B. test was Mr.

Two early auctioneers, W. L. Baird and Bob Haeger, pictured, along with Francis Darcey, founded the U.S. National Blue Ribbon Sale.

Cleveland's sale in 1910. But, if an animal reacted after reaching its new home, it was considered to be the hard luck of the buyer.

Veteran auctioneers predicted trouble when E. M. Hastings first offered a 60-day guarantee against tuberculosis in his 1913 sale. Yet, what was so controversial then, soon became a standard practice and would be expanded over the years to the publication of brucellosis test results, mastitis charts, and other health-related information about the animals. Without these better health measures, progress of purebred dairy cattle breeding would have been hindered. A joint federal-state program of test and slaughter for tuberculosis was adopted by most states in 1916 allowing indemnity payments to the owners of slaughtered animals. But it would take 30 years and countless dollars to bring T.B. under control in the dairy industry.

Late in 1914, foot and mouth disease broke out in the most serious occurrence ever, restricting public sales for several years.

When the foot and mouth quarantine was lifted in 1916, a sales boom was unleashed that would be unmatched for decades. There was a euphoria over Holsteins that made the major sales a spectacle of enthusiasm with stirring music and cheering crowds. And it was a period when sale prices took off. Caught up in the Holstein fever, some breeders took to inflating sale prices for advertising purposes ("hippodroming prices," to quote the 1930 **Holstein-Friesian History)** and many would be cheated by an active cattle trader of the time, Oliver Cabana, who soon would be accused by the Holstein Association of fraudulent practices.

With the Holstein-Friesian Association annual meetings now scheduled to be held all over the country, veteran Holstein breeder (and early-day automobile manufacturer) H. A. Moyer of Syracuse, NY, was asked to conduct the national convention sales. The Moyer national sales were held from 1916 through 1919 with great success. U.S. sales of registered Holsteins, which passed the $2 million mark in 1916, were over $7 million by 1920.

The boom and the top prices brought out a truism of the sale ring that persists today — that you can't guarantee genetics with top dollars. Here is a classic example: There was a sought-after young bull featured at the 1918 National Sale in Milwaukee called Champion Sylvia Johanna (renamed Carnation King Sylvia

Carnation King Sylvia, the $106,000 calf, is pictured with E. A. Stuart, owner of Carnation Farms; A. C. Hardy, seller; H. Lynn and Arthur Hay.

The 1920 St. Paul Sale marked the peak of the World War I boom.

immediately after the sale). During that remarkable sale, auctioneer Barney Kelley was standing on the box, frantically working to get a record price on the young bull. When someone referred to the animal's light color, Kelley replied with the classic line: "There is no milk in the hair and we don't want hair in the milk."

When the $106,000 bid of Carnation Milk Farms bought the calf, straw hats were flung into the air and then crushed underfoot in a delirious stampede over such a fabulous price. The $106,000 sale tag, which would remain unmatched in the public sale arena for over half a century, focused attention on the Carnation herd, a herd that would play a major role in the development of the Holstein breed here and abroad. Yet, the animal that caused all the excitement never really amounted to much. And in an ironic genetic twist, as King Sylvia was being purchased, a crippled and misshapen bull, King Segis 10th, was hidden in a basement cell at Carnation and not being used. He had come into the herd free with a group of his daughters and his genetics would show up in most of the great Carnation pedigrees over the years. Breeders, without the genetic tools of today, often learned the hard way that price was no guarantee of breeding power.

The boom peaked in 1920 when a total of 18,836 head were sold for over $7 million, figures not to be matched for over 30 years. That year, the first cooperative national sale, considered to be the most sensational sale ever, was held under the auspices of the Minnesota Holstein Association. In the era of big two and three-day sales, it was in a class by itself. Two California bulls from the noted herd of A. W. Morris & Sons totaled $91,000 ($50,000 and $41,000) and a Pabst cow, Pabst Korndyke Cornflower, brought $30,000 amid riotous enthusiasm. To the strains of the Star Spangled Banner, the $50,000 bull, Alcartra King Sylvia, son of Tilly Alcartra, the first 30,000-lb. cow, and by the forementioned $106,000 Carnation King Sylvia, was led into the ring. This "sale to top all sales" had 237 head from 126 consignors bringing an average of over $3000.

The war-driven inflation had fueled the sales boom, driving prices to new highs. It was the first indication of a sale ring trend — that Holstein prices through the years would tend to follow the ups and downs of the national economy, often with a lag of several years. The boom was an exciting time for Holstein sales, until the bubble burst.

The Cabana Aftermath

Many of the public sale transactions of the boom years involved Oliver Cabana, who was aggressively buying and selling animals for his Pine Grove herd. The Pine Grove herd was dispersed in 1919 at unheard of prices (up to $125,000) with 125 head being struck off. As reported in Chapter 1, some of these transactions were fictitious with the cattle never leaving the farm. Yet, at the time, the dispersal fanned the flames of the Holstein boom since it represented over 10 percent of the sales dollars that year. But, when Cabana was exposed for fraudulent testing practices and challenged by the Holstein-Friesian Association, many breeders were affected. Many herds were dispersed in the years following the Cole-Cabana scandal, often at distressed prices.

Sir Pietertje Ormsby Mercedes 37th

The fallout from the Cabana scandal, as well as general economic conditions, forced the price of Holsteins to start dropping in 1921 and continue downward through 1925. It was the severest slump the fledgling industry had ever seen, even worse than reported in the sales figures since breeders withheld the details of the worst sales.

In 1924, after three years of disheartening activity and prices, tragedy struck the breed. On April 23, the California herd of Fred Hartsook was discovered to have foot and mouth disease. Subsequent investigations traced the epidemic to garbage thrown overboard by a foreign ship off the California coast. Hartsook's herd, comprised of over 300 head representing the very tops of over 60 leading foundation herds at the time, was considered to be one of the most valuable Holstein herds in the world. The entire herd was slaughtered, including the famous sire, Sir Pietertje Ormsby Mercedes 37th (SPOM 37th), and perhaps the best known cow of the breed at the time, Tilly Alcartra. The loss not only stripped the breed of some of its finest animals, but did so at a time when the fortunes of Holstein owners were at their lowest ebb. Things had to improve.

One of the bright spots of that tragic year was a dispersal sale of the 86 Holsteins owned by John Erickson & Sons of Waupaca, WI. Built largely on the bloodlines of Sir Pietertje Ormsby Mercedes 37th, the herd drew prices that had not been seen in years. Herd members went to purchasers in 15 states, Canada, Japan, and South America, and the 86 animals became the cornerstone of breeding in 41 herds. The Erickson sale not only had far-reaching genetic impact on the breed, it also gave the industry new heart and may very well have broken the back of the depression that had plagued the industry since the Cabana case broke. Another action that would turn out to be a bright spot in the history of auction sales was the start of the Earlville (NY) sales in 1924 by R. Austin Backus, an auction series that would go on to number over 400 sales.

But the decline was not over. About the only bright spot in 1925 was the Brentwood Sale in Philadelphia where the young bull, Sir Inka May, consigned by the Minnesota Holstein Company, sold for $12,000 to Carnation Milk Farms, where he would play an important role over the next several years. First promoted and managed by W. G. Davidson and named for his farm outside

The 1924 Erickson Dispersal, which sent 86 animals to 41 herds, gave the Depression-plagued Holstein industry hope for the future.

Philadelphia, PA, the Brentwood series of annual sales, which started in 1920, fast became one of the "institutions" of the Holstein sale ring. Davidson was not only honest and able, he also had the flair of the showman and turned these sales into social events, with a special brunch for consignors and guests as well as a presale show at an open house the day before the sale. It is no wonder that "Brentwood" carried a special appeal to all breeders who ever participated. Davidson retired in 1929 and turned the operation over to his friend, Paul B. Misner, who managed the sales through 1944 when the series was ended.

The Holstein market strengthened in 1926, starting a modest recovery. The prominent sale of the year was the O. G. Clark Classic, a four-day event held in two Wisconsin locations, in which 450 head were sold. It was here that Johanna Rag Apple Pabst (JRAP), already a three-time All-American, was consigned by Joseph E. Piek of Wisconsin. He was bought by T. B. Macaulay of Mount Victoria Farms of Quebec and in the coming years, mated to the great foundation cows at Mount Victoria, would have as great an impact on Canadian and U.S. Holsteins as any sire of the days before artificial insemination.

Buyers showed new confidence in high producers as the recovery continued. At the 1927 dispersal of the Minnesota Holstein Company, the

W. G. Davidson started the famous Brentwood sale series.

A Brentwood sale scene

U.S. "butter champion," May Walker Ollie Homestead, sold for $4000 as a 12-year-old in 1927.

then U.S. butter champion, May Walker Ollie Homestead, sold for $4000 — in spite of being 12 years old. Her youngest son, Sir Bess Ormsby May, sold for $6200 to W. S. Kellogg of Osborndale Farms, and went on to become a renowned herd sire there.

The "Roaring Twenties" had been anything but roaring for Holsteins. It had been a difficult decade for public auctions, starting in scandal, sinking to new lows in a depressed agricultural economy, and slowly inching its way back to normalcy. And as the Holstein industry struggled back, breeders began to look at production records more seriously, and were willing to pay more for those animals from cows on test. During the 1925-1929 recovery, heifers from dams on long-time test brought over twice the price of heifers from untested dams. Young bulls had even a higher relative worth to buyers. This period marked an important transition point for production testing with the shift being made from AR tests on selected animals to herd tests. And, nudging progress along, the economics of the sale ring presented hard evidence that animals on sustained test were worth more at auction. But the real test was yet to come — the Great Depression.

The Great Depression

The Holstein industry was just recovering from the disastrous years of 1921-1925 when the stock market collapsed in late 1929. While the effects would take several years to materialize in agriculture, when the shock wave hit, Holstein prices plummeted to new lows. There were few public auctions during this period, and when they were held, prices were low or unreported. Breeders were suffering like the rest of the country. Widespread "dust-bowl" conditions made the situation worse. 1934 was the worst year in the 20th century for public sales: only 26 Holstein auctions were held with 1950 head selling at an average of $105.

The bright spots were few: Lashbrook Pearl Ormsby sold for $2400 at the Royal Brentwood sale in Detroit in 1931; and the next year, 7-year-old Bell Farm Rosalind, who was destined for fame as the dam of Wisconsin Admiral Burke, sold for only $180 at the Silver Glen Dispersal; and at the Third Royal Brentwood National Sale at Dunloggin in Maryland in 1934, prices were the highest in several years with Carnation Governor Prospect, son of the highest producer alive at that time, selling for $2450.

Recovery began the following year. Activity picked up although prices lagged. The Elmwood Partial Dispersal in 1936, with its $508 average for 150 head, was the best the industry had seen since the boom days of World War I. During the same year, in yet another example of how "price does not necessarily equal performance," a 5-year-old bull, Carnation Ormsby Sir Bessie, sold for $3500 at the Royal Brentwood. In spite of a rich heritage — he was sired by Sir Inka May and out of the world's record "Butter King" — Sir Bessie was never able to transmit his great bloodlines.

When the herd of Gerrit S. Miller was dispersed in 1937, after his death at age 92, an era ended. His Kriemheld herd had long been recognized as the oldest Holstein herd in North America (Chapter 1) and with names like Dowager, Agoo, and Billy Boelyn, had written a special chapter in Holstein history. Probably no other single herd up to that time had had such a great influence on the U.S. Holstein breed.

Public auction activity was picking up as agriculture pulled itself out of the Great Depression. High sale averages were again starting to occur and sales like

Weber Burke Dusty Jo was purchased for $540 at the Western Classic Sale of 1941.

the Elmwood Dispersal in late 1938 (held in a tent directly over the grave of "Queen Bessie — Queen of Elmwood," who had died three years before) brought back a taste of the old glory days. At the Elmwood sale, King Bessie Senator, unbeaten in the show ring at the time, and destined to go on to great show and All-American accomplishments, sold for $3000.

The war in Europe was heating up, but Holstein prices and activity stayed steady. Few had an inkling of the exciting days ahead in the sale ring as a nation at war clamored for high producing Holsteins, sparking a sales boom that would sweep the breed across the nation.

The Sales Boom Of World War II

As described in Chapter 2, World War II created a strong demand for milk and dairy products and brought the Holstein breed to new heights in popularity. Nowhere was this enthusiasm for Holsteins better reflected than in the sale ring, where prices immediately jumped upward with the biggest volume since 1929. Over 5000 head were sold in 1940, almost 6000 the next year, and after the U.S. entered the war, nearly 9000 Holsteins passed through public auctions in 1942.

1941 turned out to be the year of bargain buys. Dunloggin Tidy Miss went for $1000 at the Royal Brentwood to Osborndale Farms. At the same sale, 2-year-old Dunloggin Duchess Anna, who, with her family, went on to become a Michigan legend, sold for $325 to A. J. O'Connor. Perhaps the buy of the decade occurred the same year at the Western Classic, when an unknown heifer, Weber Burke Dusty Jo, was bought for $540 by F. S. Borror & Sons of California. She helped bring her sire, Wisconsin Admiral Burke Lad, into the limelight. Out of this heifer was to come one of the best known Holstein cows of the era, Rocky Hill Mont Burke Dusty Jo.

Canadian imports became a key part of U.S. Holstein history during the war years, starting an influx of registered cattle that lasted for over 30 years. These imports, brought in duty-free since they were registered in Canada, often went into commercial herds. And while many came by way of private transactions, thousands entered U.S. herds through the public auction. The Backus family, managers of the monthly Earlville, NY sales, spent countless days scouting for

C. B. Smith, shown here with Whitie Thomson, imported many Canadian Holsteins for his Wolverine Classic sales.

good Canadian cattle to "fill" their sales. Likewise, their monthly Garden Spot sales in Lancaster, PA sold many a Canadian Holstein from 1942 on. And later, when C. B. Smith began the Wolverine Classic sale in Williamson, MI, in 1946, many imports were handled. Canadian Holsteins, which would have a major impact on U.S. bloodlines during this period, came not only from the average animals, but sometimes from the top of the line.

The 1942 Mount Victoria Dispersal brought several top Holsteins to the United States. Montvic Bonheur Pietje B topped the sale at $6500, coming to Martin Buth & Sons of Michigan. Montvic Lochinvar was purchased in a joint venture by Dunloggin and Osborndale. And Montvic Rag Apple Colantha Abbekerk, an all-time great of the breed, went to Mallary Farm of Bradford, VT. However, in the eyes of some, the U.S. buyers let the best one get away — the 10-week-old calf named Montvic Rag Apple Sovereign who was purchased by Canadians and went on to greatness on both sides of the border. But late that year another widely-sought Canadian bull didn't slip away: Montvic Renown sold to the Curtiss Candy Company in Chicago for $14,100.

Holstein popularity was at an all-time high. Sale dollar volume made the biggest one-year jump ever in 1943 while the next year was the greatest in history (198 sales, 11,909 head, $3,895,039). The Dunloggin herd, owned by J. Natwick and built by the breeding skill and foresight of Paul Misner, was dispersed in 1943

Montvic Lochinvar was purchased by Dunloggin and Osborndale at the 1942 Mount Victoria Dispersal.

The Dunloggin Dispersal was one of the most famous sales in Holstein history.

The Blue Ribbon Sale in 1943 was the top consignment sale in years. Sixteen-month-old Carnation Madcap Supreme, shown here, sold for $26,000 to the Curtiss Candy Company.

in what many consider the greatest Holstein sale in history. Sale manager and auctioneer R. Austin Backus worked the box unassisted during both days of the sale and later said, "The bids came fast enough so that you could have sold them as fast as they circled the ring." Dunloggin Woodmaster's 59 offspring averaged $2588 while the complete herd of 209 head brought $383,700, the highest ever for the dispersal of a single herd. In a country sick of war news, the sale attracted attention from coast to coast, including extensive coverage in *Life* magazine. It was a prime example of how the sale ring has promoted the Holstein breed. Even more importantly, the 209 head went to 86 buyers from 21 states, spreading these famous bloodlines across the nation, and many of the animals would go on to honor in production testing and the show ring.

In spite of the war, there was still glamour in the sale ring. The 1944 Royal Brentwood Sale in Columbus, OH, the last in that famed series, brought back some of the "theater" of World War I years when a star of the breed, Montvic Bonheur Pietje B, was sold. Harvey Swartz and Vernon Hull, the two leadsmen at Brentwood, had flipped a coin to determine who would have the honor of leading this treasured female into the sale ring. When Hull brought Pietje B into the decorated ring (it was lined with ice), there was a burst of applause. Auctioneer C.

Montvic Bonheur Pietje B sold at the 1944 Brentwood for $20,000, the highest auction price for a Holstein female since 1920.

B. Smith added to the extravaganza by suggesting that when a lady like this enters, all should rise. And the large crowd had just settled back down when the bidding started without ado and in six bids and less than half a minute, ended at the highest auction price in 25 years — $20,000.

The Holstein boom continued as the World War ended and GIs came home. The influx of new ideas and energy into the industry fueled the boom into the postwar years. During the balance of the decade, several prominent herds were dispersed while consignment sales reached new highs in both participation and price. Business was brisk as the decade ended but even better days were ahead.

The Super Duper Decade

Sale ring action continued strong into the 1950's with the national convention sales leading the way. It was a period of dramatic decrease in the number of dairy cows in the nation, yet the number and percentage of registered Holsteins continued to grow. Dairymen were switching to the efficient Holstein breed, and the sale ring reflected the strong demand for good producers. The continued growth of state and county Holstein organizations (there were over 500 in the mid-1950's) brought more and more potential buyers and sellers into the industry, and local consignment sales flourished. Many local Holstein clubs began having their own sales; in states like New York and Pennsylvania, virtually every county club had a sale during the year. These local auctions, featuring cattle picked by local breeders and held at host farms, were not only a social gathering of Holstein breeders but, for many clubs, were also an important source of fund raising.

But it was not a bull market, at least for most bulls. In what was to be one of the most significant changes in the sale ring, the widespread use of artificial insemination in the 1950's killed the market for bulls, except for the highest quality animals. The full effect would take place over time, but by the mid-fifties, veteran observers had noted the change. They also noticed another change, in the names of the buyers. Now, names like Eastern Breeder's Co-op, Michigan ABA, Northeastern Pennsylvania Breeder's Co-op, American Breeders Service, and Tri-State Breeding Ring started showing up on the buyers' list. Breeders and the numerous AI firms that were springing up recognized the potential income from

The Backus sales staff in 1952

the sale of semen so the competition at the start was fierce for good bulls of breeding age. As sire proofing became more sophisticated, stud personnel focused on the female families as well as on the sires producing the top bulls. Now, sale ring prices could not easily be compared; for example, ampules of frozen semen were sometimes included in sale prices. It marked the start of the complex terms and conditions prevalent today in public auctions.

The top sale of 1950 could have been named for the decade — Super Duper. Things were super duper. The national sale figures followed the upward pattern of the late 1940's and, while dipping in the mid-1950's, held steady throughout the period. (The "Super Duper" was the name for the sale replacing the June Super Duper Sales staged annually at Earlville, NY by manager R. Austin Backus.)

The postwar boom peaked the next year in 1951, when 16,873 registered Holsteins sold at auction for an average price of $510. That mark would hold until 1967. The biggest sale event of the year was the three-day dispersal of the Sutten Oaks herd in California, with 376 head averaging $758. Exceeding the 1943 Green Meadow Dispersal, it was the Holstein breed's largest dispersal to that date. Dunloggin Strath Var, a Grand Champion many times, brought the top price for the year ($7000) in that sale, being purchased by Dr. Harold Schmidt and Edward Johnson, both of California.

In what has to be given the award for the most unusually named public auction, the 1953 dispersal of General Cochran Farm's "Uncommon Cows" in New York made the fine average of $1207 on 109 head. The herd sire, Sovereign Cochran, brought $14,000 on the bid of J. Norris Earnshaw of Pennsylvania. This was the highest sale price at an auction since 1927. The sale was one of the first efforts of a new management team which would go on to be well-recognized in the industry: R. Austin Backus, Inc., made up of the sons of R. Austin Backus — Olin, Charles, and Horace.

Farm prices dipped in the mid-1950's but Holstein sales at auction were comparatively strong. Two new sales added interest in 1955, the Crescent Beauty-Admirals Sale in Wisconsin, and the first annual Shaw's Dauntless-Dunloggin Sale in Maine. In the latter sale, all animals were from Harold J. Shaw's Shaw's Ridge herd.

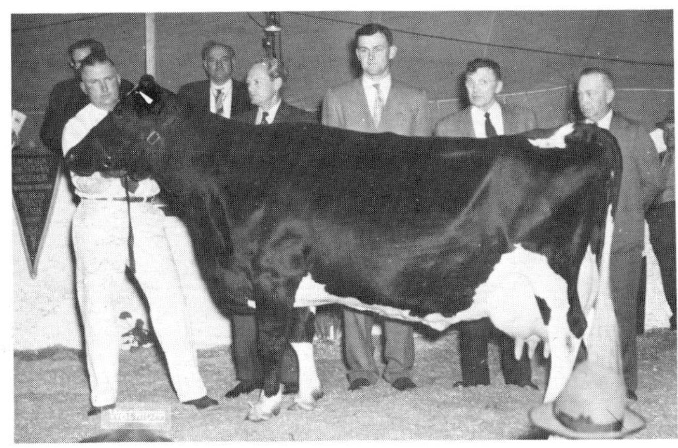

Minnow Creek Eden Delight

Harris Wilcox and R. Charles Backus

A famous production leader, Minnow Creek Eden Delight, five times above 1000 lbs. fat on 2X, was sold for $8500 as a 13-year-old at the Lakefield Farms Dispersal in 1956. Her daughter, the famed Lakefield Fobes Delight, was bought by Carnation Farms for $13,000, the top price that year. Lakefield Farms, which had been operating 40 years under the guidance of O. F. Foster, the long-time manager and Kent Mattson, assistant, sold 151 head with a $1248 average.

One of Minnesota's oldest herds was dispersed the same year, this being the herd of pioneer breeder and Holstein enthusiast, A. J. Lashbrook. His operation, the original "Home of the Pearls," was where Lashbrook Pearl Ormsby spent her earlier days.

Another important public auction series was started in New York in 1956. The Canandaigua Sales, sponsored and managed by Harris Wilcox, are a good example of the quality regional auctions that developed in the postwar years as the Holstein breed grew. Many a fine buy has been struck off at sales like this.

One of the seedstock institutions of the breed, the Osborndale herd from Derby, CT, was dispersed in May of 1957, bringing an average of $1182 for the 144 head. Wis Maestro, a senior yearling bull, sold for $30,000, the highest price since 1920. The buyer was the Michigan Artificial Breeders Co-op. In the same sale, the 12-year-old herd sire, Sir Bess Ormsby Burke Fobes (EX-GM) was purchased for $21,000 by J. S. Johnson. This was the highest price ever paid for a bull of this age, but it was due to the fact that 2200 ampules of his frozen semen

The 1950's saw AI firms entering the sale picture. Here, Wis Maestro sold for $30,000 to Michigan Artificial Breeders Co-op.

were sold with him. The days of taking a sale price at face value were over — buyer and seller had to read the fine print of the transaction.

As the decade ended, a 3-month-old calf was sold at the Whirlwind Hill Dispersal for $4000 to the First Pennsylvania Artificial Breeding Co-op. Although not much was made of it at the time, he was out of a maternal sister to another bull to gain nationwide fame — Osborndale Ivanhoe. The young bull, Whirlhill Kingpin, would go on to have a heavy influence on the U.S. Holstein breed from coast to coast.

Public auctions were different after the 1950's. On the nostalgic side, the famous families of the breed were dissipating as the genetic power of artificial insemination took hold. Herd sires were becoming a thing of the past, very few young bulls were entering the sale ring, and sharp AI firms competed for the top bulls. The tools available to buyers were crude but were improving. The herd testing program (HIR) was beginning to generate production data while the herd classification program developed type information. Soon, the power of the computer would be unleashed to guide selections. The ring had changed; few participants could just "eyeball" animals and stay competitive. But the excitement that crushed straw hats in 1918 was still there, and the auction ring, in spite of new complexities was still "where the action is."

Soaring Sales

Holstein Association Executive Secretary Robert Rumler liked to call the times "The Soaring Sixties," and there was no better place to watch the lift-off than the sale ring. AI firms were paying top dollars for top sires. Sale after sale had average prices of over $1000. Midway through the decade, observers of the auction circuit were shocked by the precedent-setting Don Augur sale, which ushered in a whole new era of down payments, extended terms, and performance guarantees. No one knew what to call the real price of the top-selling animals and strict sale-to-sale or year-to-year comparisons became fuzzier than ever. Regardless, the public sales took off during the soaring sixties on a climb that would be

Pawnee Farm Glenvue Beauty topped Lester Fishler's 1962 dispersal. The calf she was carrying, Pawnee Farm Arlinda Chief, became one of the great U.S. Holstein sires.

sustained for over 20 years. By the end of the decade, editors at the *Holstein World* would be hard pressed to find new superlatives to describe the sky-rocketing sales.

The decade started strong, but without indication of the fireworks ahead. 1970 was the Diamond Anniversary of the Holstein Association, and fittingly, the national sale was the high point of the year, with 74 head averaging $1905. The bull, Lakefield Fond Delight Fobes, who had sold for $10,000 in 1956 at the Lakefield Dispersal, was sold in 1970 for $12,800 at the Lessard Dispersal to NEPA Artificial Breeding Corporation. And Linden Dictator Wimble Wimpy, later to become the first Holstein to classify at 97 points, sold for $7000 in the Sunny Lea Dispersal to the farm's long-time manager, Nelson Rehder.

The next year a new sale called the National All-American Sale was held in Wisconsin, the opening endeavor of the management team of Harvey Swartz and Don Stouffer. And in 1962, the individual "sale of the year" came at the Pawnee Farm Dispersal when Pawnee Farm Glenvue Beauty, an Excellent 4-year-old, was struck off to Wallace N. Lindskoog of Arlinda Farms, California. She was ninth on the price list that year but turned out to be a bargain at $4300 through the son she was carrying when purchased — Pawnee Farm Arlinda Chief.

A year later two well-known herds were dispersed: Fred and Robert Nutter's Maple View herd in Maine, and the C. M. Bottema herd (Cash-Mar prefix) in Indiana. The *Holstein World* said: "The Bottema sale goes into the breed archives as one of the strongest Holstein auctions ever held." With 423 head sold in two days for $292,545, it was the largest dispersal on record.

The first sale in the Wisconsin Showcase series, "held to attract national attention to the fine quality of Wisconsin's Holsteins," took place in 1964. And it did attract national attention, gathering the highest sale average ($2281) of the year.

Another Wisconsin sale, the dispersal of Pabst Farms, had worldwide implications. Fred Pabst had died six years prior, but the dispersal reminded many in the breed of his accomplishments. Independently wealthy, he had devoted most of his life to the breeding of registered Holsteins and was an influential and dynamic figure in the affairs of the Wisconsin and National

Fred Pabst received the Premier Breeder cup from Association President Oosterhuis at the 1938 National Dairy Show.

Holstein Associations. While best known for his leadership role in the development of the first True Type models (Chapter 1), Pabst played an active role in breed affairs for half a century. Thus it seems fitting that a young bull from the 1964 Pabst Dispersal should take the Pabst name as a banner deep into the heartland of Europe and blaze new trails for U.S. Holstein genetics.

The bull was Pabst Ideal, who sold for $7000. He was selected by Dr. Gunther Rath for AI service in a West German stud now known as Rinderproduktion Niedersachen AG, located in Hanover. At the time, West Germany and most of Europe, if not using their own Friesian sires, favored Holstein sires of Canadian breeding. Ideal had a fine pedigree — pretty much a Burke with a dash of Dunloggin Woodmaster and Sir Bess Ormsby Tidy Fobes — and could transmit his high production bloodlines. Once his European-born offspring came into production, he became tremendously popular. Pabst Ideal broke down the barriers and opened the door for U.S. genetics, both sires and semen. As Dr. Rath noted in a 1983 conversation, "If Pabst Ideal had turned out differently, my career in the AI industry would have ended abruptly." It is a fascinating story and a tribute to Dr. Rath's courage and ability. Likewise, it is a fitting remembrance of the legacy of Fred Pabst.

There was another bull sold in 1964, Paclamar Astronaut, who went on to cut a wide swath in Holstein breed circles. The story of Astronaut is detailed in Chapter 13. Bred and consigned by W. R. Brooks of Paclamar Farms, Colorado, he was by Thonyma Ormsby Senator and out of one of the all-time great cows of the breed, Harborcrest Rose Milly. Just a January calf, Astronaut did not stand the trip from Colorado to Maryland too well, and his appearance at the National Convention Sale was less than impressive. Nonetheless, nine Maryland breeding establishments, along with one from South Carolina, teamed up to purchase him for $9000, which was less than many had predicted. The shaky calf went on to forge one of the finest records in the history of Holstein sires.

The average price of U.S. Holsteins at public auction jumped nearly $100 in 1966 ($398 to $494), and much of it was due to just one auction. This is how Ted Prescott, writing in the *Holstein World,* described it:

"Shock waves from the Don Augur Sale at Northford, CT, December 17th, rocked the Holstein industry all over the world with the record-breaking average of $8697 for 72 entries plus eight calves under 3 months of age. The four mature herd sires brought a total of $316,000, two of them selling for the all-time high auction sale price of $108,000 [each], and all four going to American Breeders Service, DeForest, WI."

Indeed, the sale prices did produce shock waves throughout the Holstein industry as a whole new way of buying and selling cattle, extended terms, became more common. And the Augur sale was not the only splashy one — the National Convention Sale in Kansas had a $3236 average. For the year, the *Holstein World* reported 18,990 registered Holsteins sold at an average of nearly $500.

While there was some controversy generated by the "promotional prices" paid at the Augur sale, it also became soon apparent that the high prices for animals of great genetic potential were justified by the tremendous earning potential of the animals. The sale was the first one reported where the sale terms on specific animals were publicly stated. Terms called for a down payment, insurance against death of the bull, and regular payment based on semen sales. The four top herd sires, for which ABS had paid $316,000, were paid for out of semen sales within 18 months. And while this new sale environment made it difficult to accurately

Paclamar Bootmaker sold for $80,000 in 1967.

report the cost of animals at auction, it was just a glimpse at the complicated purchasing arrangements of the 1970's and 1980's. By then, it would be rare to find a top animal bought in a straight purchase. As elsewhere in business, the days of "cash on the barrel head" had vanished.

Public sale prices started up in 1967 on a climb that would continue into 1985. The Paclamar Farms Sale drew people from great distances and brought the year's best price, $80,000, for Paclamar Bootmaker, the son of Snowboots Wis Milky Way. The next year, the highest price for a female, $44,000, was paid at the Tara Hills Dispersal for Future Hope Reflector Blacky.

As the decade ended, the National Convention Fiesta Sale in Anaheim, CA, set an all-time high average for U.S. consignment sales ($4392 on 50 head). And one of the oldest remaining herds in the country, the Winterthur herd in Delaware, the result of 55 years of constructive breeding by owner Henry F. du Pont, was dispersed in 1969. Mr. du Pont, who had taken the herd over from his father in 1914, had bought the Spring Brook Bess Burke 2d family (five members) in 1918, forming the backbone of the Winterthur breeding program for the next half century. Mr. du Pont's death at age 88 earlier in the year had led to the dispersal of 135 head, all homebred in maternal line for up to a dozen generations. An important era in Holstein history had ended.

It was far from the end of an era in the sale ring; things were just beginning to get exciting. The 1960's had brought back a taste of the old days, as record after record was broken. But there was no time to stop and catch a breath. The next decade, with embryo transfers and the first chunks of investor money showing up, would bring an entirely new picture to the public auction.

Tara-Hills Pride Lucky Barb set a world's record auction price of $122,000 in the 1972 Hanover Hill Sale.

A "Bull Market" For Holsteins

The sale ring, ever the sensitive barometer of the industry, rose with the swelling confidence in Holsteins in the 1970's. As pointed out in the Foreword, breeders had begun to use the new evaluation tools available to them in the late 1960's, and at once, Holstein genetic progress shot upward. So did public auction prices — sharp breeders were willing to pay top prices for top genetics. As breeding was changing from an "art" to a "science," so did cattle buying. Terms like "predicted difference" and "total performance index" entered sale ring terminology. Later in the decade, when syndicates were more active and outside investors started participating in sales, pre-auction conversations were more likely to be sprinkled with expressions like "cash flow," "accelerated depreciation," and "investment credit." Investors became interested in Holsteins at the time that the technology of embryo transfers was emerging and it was love at first sight. Just like artificial insemination had done years before, ET turned the auction price structure topsy-turvy.

The adage "Price does not determine quality; rather, quality determines price" held true in the 1970's. The Holsteins offered at public auction were the finest the breed had seen: high-producing animals with carefully selected type traits. It was a decade of steady growth of the Holstein breed, both here and abroad, and sale ring prices and averages just kept climbing. It was a 10-year "bull market" for male and female registered Holsteins.

As the decade began, the World Premiere at Madison set a new all-time average ($5480) for consignment sales for the second year in a row. The top price of the year ($40,000 for Gray View Coral Shamrock) was set there. There were also several important dispersal sales headed by the late Edward M. Boehm's Duncravin herd in New Jersey and the long-established Mallary Farm herd in Vermont, founded by DeWitt and Gertrude Mallary. And the largest herd ever sold at once (573 registered Holsteins), Diamond J Holsteins owned by Edward Johnson, was dispersed in California.

More and more syndicates were showing up in the public auction buyers' lists, particularly for the top animals. Consisting primarily of breeders and similar to those operated in real estate and other business ventures, syndicates brought any

Hanover-Hill Triple Threat-Red drew the highest price ever paid for a Red & White Holstein.

number of breeders into an organization set up to purchase one or more animals. In the eyes of many in the business, syndicates tended to force the price of certain animals up to artificially high levels. Later, when outside investors joined breeders in these organizations, this complaint would be voiced by more and more farmer/breeders.

The market was strong. Fed by a period of inflation and relatively stable milk prices, Holstein prices swung higher. In reporting the sales of 1972, the *Holstein World* became positively lyrical. No superlatives seemed adequate to describe the year. Every public auction record — average, total volume, number sold —was broken. And the next year was even bigger. One sale that drew a tremendous crowd in 1972 was the Hanover Hill Sale where the world's record price for a dairy cow was set. Tara-Hills Pride Lucky Barb did not need "luck" to earn her $122,000 price tag paid by a Canadian syndicate. Her red & white son, Hanover-Hill Triple Threat-Red, selling for $60,000 to ABS, set a world's record for a Red & White Holstein. All in all, eight members of her family sold that day for a total of $350,000.

While syndicates were playing a role in high prices for top cattle, another group was becoming more and more active in the auction ranks. The Holstein Association had expanded its network of consultants and was becoming more active in helping members find markets for their cattle. State associations were doing likewise, having more sales of their own. County club sales dropped in number as more and more district level sales were being held.

In 91 years of sale reporting, 1973 was the first year to hit $20 million in sales. It was also the first time that the auction average reached the $1000 level. Helping set the record was the Hall of Fame Sale in Pennsylvania, held to commemorate the 60th anniversary of the founding of the Backus Pedigree Company in Mexico, NY, by R. Austin Backus when he was 16 years old. The sale, the first in which every animal's price exceeded $2000, also had the highest average to date ($7197) for any dairy breed consignment sale.

Carnation Farms had their first sale since being founded in 1910 and sold over half of the 67 head to buyers from Canada, Italy, Japan, Mexico, and West Germany. This auction, taking place at a time when exports were growing,

Gene-Acres Felicia May was purchased by a 22-member syndicate.

pointed out the demand for U.S. Holsteins overseas, a demand from which many breeders in future years would profit.

One of the original investor groups, Leadfield Associates, was the buyer of the top-priced animal of 1973 when they paid $70,000 for Md-Maple-Lawn Marquis Glamour in the Allen Dairy Sale of Champions in Pennsylvania. Leadfield Associates was a group of investors who, with John J. Sullivan of Batavia, NY as leader, had established a fine Holstein herd in the early 1970's. Spurred by the profit-potential of the Holstein breed, they were the vanguard of a wave of outside investors who would have a marked effect on sale prices of the future.

The Crescent Beauty Admiral Sale was planned before the untimely death of Allen Hetts and was carried through by his wife, Doris, and family. Held at the Hetts farm in Wisconsin, it was attended by an overwhelming crowd and was a fitting tribute to Allen Hetts. It also provides an excellent example of how syndicated purchasing worked in the mid-seventies. Featured at the sale was Gene-Acres Felicia May Fury (2E-97-1215 fat 3x), at that time once an All-American and four times Reserve. She was purchased for $80,000 by a group aptly named The Felicia May Syndicate which consisted of 22 members from seven states scattered across the country.

Pomp and circumstance entered the sale ring more and more as the decade progressed. The two top sales of 1975, the Royal Erinwood, which averaged $19,304 on 24 head, and the Designer Fashion Sale, averaging $24,053 on 38 head, were held in New York state. The Designer Fashion Sale, planned by the Hanover Hill Sales & Service team of David Younger, R. Peter Heffering, and Stephen Briggs, set records that would not be exceeded for another five years.

The price rocket had gone into booster stage. By 1976, for the first time in history, more than half of the reported auction sales averaged over $1000. The year before it had been 40 percent, the year before that 18 percent. A new sale, the Sunny Knolls Invitational with host F. John Holmes, was held in central New York and achieved the best average for the year ($9825). Other top sales were the Royal Erinwood, the National Convention Sale, the World Premiere, Wintercrest Invitational, RuAnn Fiesta, Curtiss Classic, and California Premiere. The state association sales were coming into their own, Minnesota

The Designer Fashion sales of the late 1970's set records for consignment sales.

topping the list of 32 with the 2nd Golden Gopher Classic.

Prices continued to escalate in 1978 with six sales averaging above $6000 a head. The Winter Place Sale included a practice which, while not completely new at the time, has become much more prevalent in sales of the 1980's. Substantial contract offers on future heifer calves were offered from a dozen or more of the top animals sold. At that sale, Miss Milliehoe, yearling daughter of Osborndale Ivanhoe, was purchased by a syndicate for $80,000, and Winter Place owner C. J. Auger offered a $40,000 contract for the calf she was carrying, if a heifer.

As the decade ended, the MGM Grand Sale was held in the ornate setting of the MGM Grand Hotel in Reno, NV. The only sure bet in the house was registered Holsteins. Smart breeders knew that and proved it with an $8042 sale average. Yet the era that had seen record after record broken had one more superlative left. Mike and Marion Dolan's Drifty Farms Dispersal in California set an all-time record for U.S. Holstein sales volume, when 712 head plus 25 calves sold for just under $2 million.

A song from "Oklahoma" says, "They've gone about as far as they can go." But had Holstein sale prices? As the decade ended, auction prices seemed poised. Would they level off or would this new investor money drive them higher? The answer would not be long in coming.

New Times — New Ideas

There has often been a certain "theater" aspect to the public sale ring. Whether the 1918 sale with hundreds of boaters flying into the air in celebration, J. E. Mack performing magic tricks in the auction box, a 1944 sale ring surrounded by crushed ice and admiring buyers, or a 1979 sale in the lobby of the MGM Grand Hotel, the pomp and circumstance of top Holstein sales have made them the glamour events of the industry. As the Holstein Association approached its 100th year of operation, the sale ring still had the excitement of the boom years, and entrepreneurs were making New Year's resolutions on how to better promote Holsteins. The ideas they would come up with in the early 1980's to auction cattle would shock some and beguile others, as they competed with the thoroughbred industry for exciting sales and exotic settings. While the bulk of registered Holsteins sold at auction were at the popular county and regional sales, the major sales often featured not just top animals but complex genetic packages and extended terms. Whether at a Texas barbecue or an Hawaiian luau, and whether to the strains of a jazz ensemble or a steel band, the presale and postsale events were memorable. So were the settings: resorts in the Catskills, Madison Square Garden, and Waikiki Beach, to name a few.

There were changes in the auction market. Many top Holsteins would be auctioned off before they were born. The breed's elite animals would probably never see a dusty sale ring. The days of a farmer/breeder walking into a top sale and bidding cash for an animal had all but ceased. Major auctions lost much of their "public sale" nature due to the complex financing alternatives associated with top animals. You just could not put together a deal at ringside: attorneys had to be consulted, tax angles considered, and often, sellers retained partial interest in the animals being offered. Most of the deals were put together beforehand.

The auction market became two-tiered in the 1980's: there were the elite sales, with mostly investor funds; and there was the more traditional farmer/breeder market. The prices of Identified Holstein females and average to good registered Holsteins tended to follow the supply and demand cycles of the past, often rising and falling with the beef market. The genetically superior animals just kept rising in value. Outside investors, accustomed to reading financial reports and stock columns, zeroed in on the high production index animals, concentrating on PDM and PD$, in most part caring less about type. The prices brought by top genetic animals, animals with the high indexes and pedigrees of promise, surprised even expert sale managers.

The one sale that perhaps best captured the flair of top Holstein auctions in the 1980's was the unique Holiday ET Extravaganza held on New Year's Eve, 1982, in the Bahamas. The first auction sale ever to take place under U.S. management outside continental North America, the event was the brainchild of Ed Fellers of Kansas, who, in partnership with another Kansan, Dallas Burton, staged the event. The report of the sale written by the *Holstein World's* Mike Young said: "...this sale offering was a breed first as no consignments were present... Most, in fact, were not even born but the 58 lots averaged $13,599 on the strength of the cow families. Selling was a choice of ET's (embryos), male and female from the top line pedigrees offered in the stylish catalog...

"The sale offered no barn crew, no leaders and trucking was no problem. The key to sale night was Danny Weaver and Pat Jacobson from Agri-Graphics [Illinois] who had slides of the maternal families cleverly arranged to fit the call of the pedigree reader... and had 16 x 24 color pictures of the dams ready for the buyers to take home..."

The first Holiday ET Extravaganza was held on New Year's Eve, 1982, in the Bahamas.

The 1980 National Convention Sale set the stage for the centennial decade with its "Stars of the Stage" sale held at the Grand Ole Opry in Nashville which became the highest average ($10,465 on 117 head) and the highest volume for a consignment sale ($1,230,500) ever held.

The Hilltop Hanover Sale, held later that year in New York and consisting solely of animals from their farm, set new records for prices and volume. An investor group, Dreamstreet Holsteins, paid the then top price of $300,000 for Pammies Citation Paula-ET, the eighth generation Excellent from the family that goes back to Future Hope Reflector Blacky, who herself was a top seller in 1968 at the Tara Hills Dispersal.

Other top sales of 1980 were the Maine State Sale, RuAnn Fiesta (R. A. & Douglas Maddox) and Summer Dreams by Dreamstreet. Dispersals in 1980 included Dr. David Smokler's (soon to be Holstein Association President) milking herd, Tom and Sandy Morris' Deronda, Jack and Emra Stookey's, and Morris Brothers Arwyn Dispersal.

Sale prices continued upward. The 1981 Designer Fashion Sale averaged $44,891 on 46 Holsteins. Other top sales included the Gold Rush Sale in California (Alvin and Jim Quist), the Poverty Hollow Sale in New York, and the Great Expectations Sale in Pennsylvania. Curtiss Breeding Service, a leading AI firm, folded during the year and sold 38 bulls at auction. The number of animals sold in 1981 (22,902) and the sale volume ($66,109,195) are record figures as this history goes to print.

1982 — the year when the world found out that there is such a thing as a million-dollar cow — blew away all past notions on what top Holsteins would bring at auction. Fueled by outside investor money and the promise of elite genetics through embryo transfer and still operating under favorable tax regulations, the public auction market exceeded all previous high-water marks.

Allendairy Glamourous Ivy (2E-96-GMD) became the first million-dollar cow in history at Thomas Pearson's Pearmont Dispersal in Wisconsin when she sold for $1,025,000. After the winning bid by Cormdale Farms, Inc., the crowd spontaneously rose as one, clapping and cheering in celebration of this historic event.

Allendairy Glamourous Ivy was the breed's first million-dollar cow.

While the "million-dollar cow" captured the fancy of the public, the top animals were catching the attention of more and more outside investors who, in most cases, were more familiar with the other "stock market." With expert help, they made purchases that resulted in several six-figure sale averages for the year. Reporting the first Holstein sale to reach a $100,000 average, the *Holstein World* reported:

"With a buyer's list that reads more like the *Wall Street Journal* than the *Holstein World,* Dreamstreet Holsteins, Inc., NY, had an impressive sale on August 15 [1982]." Only 16 head were offered, all from Dreamstreet herds, selling for an average of $106,812.

The other six-figure sale also involved an investor operation, Odyssey Farm. The Odyssey Elite Sale, held in New York, brought $825,000 for Hanover Hill Sheik Barb-ET although the owner, Hanover Hill Holsteins, retained an interest. The entire sale of 33 head averaged $118,106.

While 1982 was a banner year for top Holsteins at public sale, uncertainties about milk-pricing policies and a widely publicized milk surplus affected the industry the following year. Still, the market for the elite Holstein was as strong as ever. The average sale price rose to a new record of $2904 for 357 cataloged sales.

The Designer Fashion '83 Sale held in New York City under the management of Hanover Hills Sales & Service set a new high sale average of $145,387 for 62 head. At the same sale, a new record price for a single bull was set when Dreamstreet Sexation Diligent, the son of Ocean-View Sexation from an EX-93 Arlinda Chief, was struck off at $610,000.

The high sales volume, the high prices, and the influx of new buyers into the industry raised concerns about buyers failing to pay for animals purchased, or sellers not living up to their obligations. Ethical standards for Holstein sales had been a written policy of the Holstein Association since the 1948 convention but it was time to re-emphasize certain issues. A restructured merchandising policy, covering public as well as private sales, was made effective in July 1982 and amended in 1984. Aimed at creating a healthy, honest, ethical sales environment, this "ethics in sales" initiative of the Association was strengthened further in 1984 by the assignment of Charles Detch, a seasoned Holstein Association staffer, to

A feature of the Southfork Sale was Cross-Country Sunshine Nugget (VG-88) who sold with six embryo pregnancies for $125,000.

the task of building public support for ethical sales of Holstein cattle.

1984, the last year of the Holstein Association's first century, saw the market for top genetics continue to be strong. Several glamour events symbolized the Holstein's rise to the star status once reserved for Kentucky yearlings. The last big sale of the year epitomized the exotic settings that had characterized the top sales in past years. Held at the site of the popular "Dallas" television show, the Southfork Ranch, the sale featured fancy advertisements, fancy settings, fancy pedigrees, and as might be expected, fancy prices. It was the largest dollar grossing sale, at $11.9 million for a consignment sale. The $1 million price for one-half interest in Paclamar Tidal Wave was the highest recorded price for half interest in a bull in a public sale. It was also the first sale in Holstein history where a cow that has made in excess of 51,000 lbs. of milk was sold. Long-Haven Wayne Stephanie *RT (EX-90) sold with nine unborn embryos out of her and her maternal family for $1.1 million. She was the highest producing cow to be sold to date.

The Southfork Sale, with its 41 lots of donor cows with born and unborn offspring, showed the emphasis of marketing high-dollar cows to the investor market. It had all of the "theater" effects of the long-remembered sales of the past and, while not typical of the hundreds of public auctions held each year all across the country, provides a fitting end to the public sales covered in this book. For as the Holstein Association begins its second century, the sale ring is perched on another brink. Will outside dollars continue to influence Holstein prices, or will changes in tax laws or rulings bring the market, in the eyes of some, back to reality?

The subject is too current to have the advantage of history's hindsight, yet it is important to note that the impact of outside investor money has been a factor, and a controversial one, in the developments of the last 10 years in the sale ring. The public sales of the 1980's were influenced by a phenomenon created by favorable income tax laws. While moneyed individuals like Fred Pabst and Henry du Pont had been influential Holstein breeders in the past, they were also farm owners. Now Wall Street attorneys, bankers, rock stars, and professional athletes were drawn by income tax advantages into the ranks of registered

Holstein owners. Some never saw the animals they owned. The money pumped into sales by outside investors elevated prices to a point where most farmer/breeders could not compete for the top genetics at open sale. Feeling priced out of the market, some breeders decided to join in the action, and housed investor-owned animals with their own herds. Others were content to take the high prices offered by the outside money, recognizing it as a situation that would abruptly change with a change in the income tax laws. Perhaps the feeling of the typical farmer/breeder of the 1980's was described by Richard Coyne, a Holstein Association board member, who said, "If I'm going to get up at 2 o'clock in the morning with a cow that's calving, it's going to be *my* cow."

Zane Akins, Executive Secretary of the Holstein Association, put the investor phenomenon in perspective. He said, "The money they brought in went to people who depended upon it for their livelihood. They bought the good cows and paid a little more money than the farmer/breeders could get elsewhere. They researched, they went to the sales, they got some excitement going. But they're in it for tax reasons, they're not in it for the long term. They have no commitment to stay in for 20 years or a lifetime. It's a passing fad."

Only time will tell the outcome. In any event, since the 1800's, the sale ring has attracted new energy and money to Holsteins: breeders who switched from other breeds in the 1920's, GIs returning with "mustering-out" pay in their pockets in the 1940's, and suave investors with geneticists picking their animals in the 1980's. For over a century, the Holstein cow has demonstrated that she is a good investment. It is this "backing a winner" feeling that has made the sale ring such an exciting part of the breed's rise to prominence.

chapter 9

The Show Ring

The show ring is where many Holstein breeders — as they watch the 4-H calf they have worked so hard to prepare for this moment — first feel the satisfaction of raising a fine dairy animal. There, in front of a cattle judge and a crowd, all the work and effort comes into focus. The show may be a "Black & White" show at a county fair, but to the exhibitor, it feels like the National at Madison. And so it always has. The excitement, the sense of accomplishment that comes from success in the show ring is worth more to most exhibitors than a wall full of blue ribbons. Whether at a "Black & White" show in a rural western county, or a national show with a world-known judge, the enthusiasm, camaraderie, and competition of the show ring still draws exhibitors, spectators, and top Holsteins, just as it has done for over a century.

Throughout the past 100 years, some have scoffingly called the show ring anachronistic, obsolete, no longer meaningful, and yet, Holstein shows are still around, strong and vital. The Holsteins winning the top classes and being named All-Americans are the best ever: high producers with top classification scores possessing that "dairy" look that breeders and judges like.

During 1984, the *Holstein World* reported on 84 major Holstein shows in 46 states where 14,789 registered Holsteins went through the ring. There were many smaller shows that went unreported. To the surprise of some of the detractors of the show ring, enthusiasm and participation remain undiminished. Shows are still an important marketing opportunity for the serious Holstein breeder.

The thrill of the show ring affects young and old.

Since the turn of the century, Holstein shows have given "ringside jockeys" a chance to view the breed's top animals.

The Early Years

From the earliest days of Holsteins in the United States, the show ring has been recognized as an important factor in the breed's progress and development. At the outset, when animals of this breed were something of a novelty, the leading breeders needed the advertising value of such displays. As early as 1864, Winthrop Chenery (Chapter 1) exhibited cattle in New England, and by 1873, several showings of Holsteins were made at New England and New York fairs. That year and the next, Gerrit S. Miller of Peterboro, NY, exhibited his 1869 imports and their offspring at the New York State Fair, competing with the successful New England exhibitor, W. A. Russell, and others.

It was not until five years later, however, that competition became really keen. Smiths & Powell Co., Syracuse, NY, started showing locally in 1876. By 1885, they were the largest concern connected with the industry, maintaining a herd of 500 head and importing as many as 400 head in a year. While their showing was limited principally to New York, they had a great record of winnings for many years.

The earliest shows were mainly held to help sell cattle, to bring the practically unknown Holstein breed to the attention of prospective customers of importers and breeders. As the 19th century progressed and Holsteins became more prevalent, the show ring provided other important advantages. Among these were (1) the setting of standards for desirable conformation (type) agreeable to the majority of breeders; (2) opportunities to see as well as show the best representatives of the breed; (3) a setting where breeders could exchange experiences and opinions; and (4) stimulating opportunities for competition. Then, as now, the show ring provided successful exhibitors with one of the finest merchandising tools available to any breeder.

In the late 1800's, exhibitors faced difficult travel conditions and high expenses but still did a surprising amount of showing. A number of important shows were held and the winning animals received wide acclaim. The noted show bull, Billy Boelyn, imported by Gerrit Miller in 1876 was one of the early stars of the show ring. He was Champion at the New York State Fair in 1880 and later, in the herd of Edgar Huidekoper of Pennsylvania, added an impressive list of other winnings. His sons went on to head many prominent herds.

Another example is Netherland Prince, Smiths & Powell's great winner in the years 1881 through 1884. Smiths & Powell, like Mr. Miller, combined their show ring activities with liberal and continuous effort to demonstrate the superiority of the Holstein breed in production. As a result, Billy Boelyn and Netherland Prince are perhaps the two great early progenitors of the Holstein breed in America.

The first national showing of Holsteins was at World's Fairs, the World's Columbian Exposition at Chicago in 1893 being the first. Thomas B. Wales,

Sarcastic Lad and Jolie Johanna were Grand Champion male and female at the 1904 St. Louis World's Fair.

The Wisconsin State Herd at the 1924 National Dairy Show featured Johanna Rag Apple Pabst (left).

secretary of the Holstein-Friesian Association, was judge. Exhibitors came from New York, Kansas, Michigan and Canada.

Much of the early show enthusiasm developed in the nation's heartland. Through the activity of several well-known breeders of the time in Iowa, Missouri, Minnesota, and Wisconsin, the show rings of the Middle West attained a high standard during the closing years of the 19th century. Adding interest and competition, too, were herds from outside the area, notably that of Henry Stevens & Sons of New York State, who showed with considerable success at the Illinois and Wisconsin State Fairs and the Canadian National (1897).

Other shows of the period attracting national attention included the Pan-American Exposition at Buffalo, NY, 1901, and the St. Louis World's Fair of 1904. At the latter show, the male Champion was Sarcastic Lad, recognized as one of the great transmitting sires of the breed's early years. It was perhaps the greatest of his sons, Colantha Johanna Lad, who later achieved recognition as the grandsire of the dam of Wisconsin Admiral Burke Lad (Chapter 11).

The last World's Fair to have an influential Holstein show was the 1926 Sesqui-Centennial at Philadelphia. Here, 156 head were shown by 19 exhibitors from 12 states. Triune Papoose Piebe was Junior Champion as a calf for Hargrove & Arnold of Iowa, her breeder and exhibitor. She was All-American that year and the next, after which she was sold at auction for $11,100 to Mount Victoria Farms in Canada. She became one of their most valued foundation cows and the only female ever to win All-American at every age from calf up through maturity.

The National Shows Before World War II

While the early World's Fairs helped bring Holsteins to the attention of the public, national shows have been much more important in the development of the breed. Likewise, state, regional, and county shows have made much more of an impact on the Holstein industry in later years.

The first National Dairy Show, held in Chicago in 1906, was the idea of one of the breed's pioneers, W. B. Barney of Home Farm, Iowa. Barney not only led the drive to hold the show, he also was instrumental in establishing its sponsor, the National Dairy Association. He was a successful exhibitor at the early nationals as were others who would help make Holstein history: R. E. Haeger, W. S. Moscrip, and Frank White. Judges included names like Clarence F. Hunt, F. H. Scribner, W. J. Gillett, and Ward W. Stevens.

The National was primarily a midwest event in the early years at Chicago. The first invasion of Pacific Coast breeders occurred in 1912 when Hazelwood Holstein Farm of Washington came east and returned with 13 out of 18 first places. Two years later, a disastrous outbreak of foot and mouth disease forced the animals into extended quarantine and cancelled the show the following year. It seemed advisable to get out of the Chicago stockyards, so in 1916, the National became a road show, remaining so until World War II.

The 1916 show, held at Springfield, MA, began one of the important features

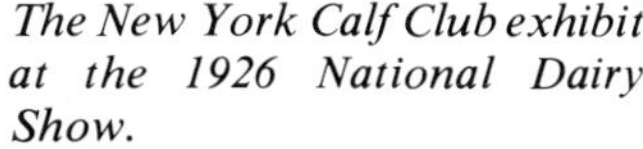

The New York Calf Club exhibit at the 1926 National Dairy Show.

of nationals, the state herd contest. Three years later, back in Chicago, the Minnesota Holstein-Friesian Association took the concept one step further, initiating the idea of cooperative showing. By sponsoring animals from many herds, they allowed many more Holstein breeders to participate. The National immediately took on a more populist flavor: the following year, 200 head, with 90 different owners, were shown — a new record for all shows of this date. The state herd contest was adopted as a permanent project.

The idea of cooperative showing, which would later include county herds at state fairs, could not have come at a better time. After AR testing became popular, breeders had lost interest in type traits. During the boom days of World War I, production records were the rage and the show ring faded in popularity. Many active breeders felt that showing was not only a waste of time, but that it hindered record-making. However, when the Cole-Cabana scandal broke with the accusations of falsified records (Chapter 1), that, coupled with a general depression in the industry, provoked a reaction against record-making and triggered one of the worst eras in Holstein history. For better than a decade, the show ring was one of the few morale-builders that Holstein breeders could cling to.

At one time it had been hoped that St. Louis, with new facilities built especially for the purpose, would become the permanent home of the National, but this dream died with the Depression.

After a two-year suspension due to the Depression, St. Louis again was the site of the National, but lack of interest there put the show back on the road. Up until the outbreak of World War II, there were several successful nationals in Columbus, OH, as well the first visit to the West Coast, in 1939, as part of the World's Fair. Prominent exhibitors of the era included Carnation Milk Farms, Pabst, Elmwood Farms, Maytag Dairy Farms, and Franlo Farms. W. S. Moscrip continued a long tenure as national judge until 1938, when Paul B. Misner made his debut.

With World War II looming, the National Dairy Show was discontinued after the 1941 Memphis show, and the National Dairy Association folded soon thereafter. The National had inspired the construction of adequate show facilities

Paul B. Misner, one of the breed's foremost breeders, made his National Dairy Show judging debut in 1938.

at every location away from its original Chicago home. Most importantly, the show had made a great contribution to the Holstein breed during a period when the show ring was where the standard for functional type traits was set.

The Dairy Cattle Congress

The second great national annual dairy show to be established was the Dairy Cattle Congress at Waterloo, IA. Older members of the dairy cattle scene still recall with nostalgia the long series of "Waterloo" National Shows.

The show had a long history, and, as with the earlier National Dairy Show, veteran Iowa breeder W. B. Barney was instrumental in its founding. Starting as a display of purebred dairy cattle in conjunction with the Iowa State Dairy Association meetings (of which Mr. Barney was an officer), the idea became so popular that when the Iowa Dairy Association met at Waterloo in 1910, a competitive show was held, with prize money offered. It continued annually as an all-breed show with increasing participation. In later years it became known as the National Dairy Cattle Congress, as it was truly a national competition featuring the leading show herds and showmen of the United States, as well as Canada.

The derivation of the name "Dairy Cattle Congress" has been described by Ed Estel, secretary-manager of the show from 1915 until his retirement in 1958. About 1912, a group of the officers were discussing the need for a distinctive name. The talk drifted to national politics, and someone remarked that these good purebred cows had more sense than many congressmen. Wilbur Marsh, a noted Guernsey breeder, remarked, "That's it—let's call our show 'The Iowa Dairy Cattle Congress.' "

The Waterloo show functioned as the National Holstein Show during the Depression years of 1932-33-34, sustaining Holstein interest during a crucial time in the industry's history. The same was true for 1942, after a permanent closing of the National Dairy Show.

There were few shows of more than local interest during the war years of 1943-44-45. With the resumption of showing in 1946, Waterloo became the officially designated National Holstein Show. The competition which followed

The Waterloo show was "old home week" for Holstein enthusiasts for many years.

established the reputation of many Holstein breeders and wrote a unique chapter in the Holstein show ring. During its banner years, Waterloo was the "old home week" of the entire show season, bringing together the best cattle of all dairy breeds, the top showmen in each of the respective breeds, and interested owners from across the country. It was an all-breed congregation with the inter-breed rivalry second only to the opportunity for all to get together with friends, old and new, joined by the common bond of appreciation for great dairy cattle. It lasted through 1965.

In November of that year, officers of the dairy breed associations were notified that the 1966 Waterloo Show was scheduled for July, a date unacceptable to the dairy breeders, being too early for a national show. After dairy breed officials registered their displeasure, Dairy Cattle Congress Manager Maurice Telleen scheduled an inter-breed meeting for November 30, 1965, at Chicago.

Representing Holsteins at this meeting were Holstein Association Director A. C. Thomson, Association Secretary Robert H. Rumler, and F. C. Barney, long-time *Holstein World* associate editor.

It was apparent at the meeting that no officials of the Congress were particularly concerned with purebred dairy cattle nor with the dairy industry as a whole, despite the financial success of the show. The Congress offered a few concessions in dates but also wanted to decrease the number of classes to be shown.

A month later, at the December, 1965, meeting of the board of the Holstein Association, it was voted unanimously to withhold Association recognition from the 1966 Dairy Cattle Congress. The "Waterloo" era was over.

International Dairy Show

In 1953, the first International Dairy Show was held at the Chicago International Amphitheatre under the same general sponsorship and management as the long-established International Livestock Exposition. The dairy show at Chicago was an immediate success in numbers, quality, and breadth of competition; it was on the same level with the National Show in Waterloo, at least in the Holstein breed, until the latter lost its "National" designation.

At first, the Chicago show was held in early October, reaching a peak in 1958 when 403 Holsteins paraded through the ring before Judge Hilton Boynton of the University of New Hampshire. Starting in 1959, the date of the dairy show was moved to later in the year to coincide with the dates of the International Livestock Show. This later date had less appeal for the majority of dairy exhibitors, but entries still arrived from coast to coast—and frequently in considerable numbers from Canada.

In 1966, when the Holstein Association did not designate any show as "National," it was generally felt that the Chicago show filled that position, at least unofficially. That year, 349 Holsteins were brought to Chicago by exhibitors from 21 states and Canada to be placed by Judge John W. McKitrick of Columbus, OH.

The Chicago International Dairy Show continued through the 1969 season. In early 1970, Judge Robert L. Hunter, Holstein breeder and president of the International Dairy Show, announced that the show and affiliated dairy events would be combined with the World Dairy Expo (the Central National Holstein Show) at Madison, WI.

In spite of its big-city location, the International's 17-year existence, from 1953

The International Dairy Show was held at the venerable Chicago International Amphitheatre.

Executive Secretary Rumler, shown here presenting the Premier Exhibitor award to Peter Heffering at the 1963 Eastern States Exposition, was instrumental in making the change to a three-national-show system.

through 1969, found 5031 registered Holsteins exhibited in the individual classes. Quality was high; the management, officers and directors, some of whom were Holstein breeders, always had the purebred interest at heart. It was truly an international show and winning carried its full quota of prestige. But, times and conditions changed and the decision was made to move the show to the World Dairy Expo.

The Three National Show Concept

Because of the late notice given the Holstein officers of the changes in date and format for the 1966 Waterloo show, the Holstein Board of Directors refrained from naming a National Show for 1966. Instead, the board used the interim period to study the entire question of the national show concept.

Association Director of Extension James F. Pound made an exhaustive survey of the major shows of recent years, researching the number of animals shown, exhibitors, state herds, and spectators. Based on this survey, Pound and Association Executive Secretary Robert H. Rumler came up with an innovative—some considered it radical—approach to the national show which was presented to the Extension Committee. They proposed that not one, but three national shows—Eastern, Central and Western—be held annually.

It was a move in concert with the movement of the Holstein industry, which, at the time, was shedding some of its trappings of traditionalism. Color marking requirements were being loosened, and Red & White Holsteins were being accepted into the herd book. The attitudes that were bringing these changes to the registered Holstein scene were also being felt on the show circuit.

The concept of three national shows arose out of this changing environment. It had been costly to move state herds from all across the country to the Midwest for the one national show. Breeders in the East and in the far West felt, with some justification, that they didn't have a fair shot at the Waterloo show, since exhibitors from the Midwest had the opportunity to get more cattle there and dominate the classes.

The concept was to de-emphasize the excessive amount of attention being given to relatively few cows — to put the show program in its proper perspective. The three national shows would be equal, allowing more animals to qualify and thereby creating a broader representation. It was another move by the Holstein Association to reach out to more breeders.

After considerable discussion, the extension committee adopted the Rumler-Pound concept and brought it before the full board on May 29. The board, in turn, while not completely unanimous, approved the recommendation, and the idea was presented as an accomplished fact at the 81st annual convention of the Holstein-Friesian Association of America at Wichita, KS, on May 31, 1966.

Extension Committee Chairman, Dr. Sims explained it to the Convention: "The status of the show — national, regional or local — has little effect on the size of the ringside and, to a lesser degree, the number of cattle exhibited. The regional shows reveal a truer across-the-board participation from the areas they serve when we compare the number of state herds shown to the number of states represented. With this knowledge, a realistic appraisal of the time and an appreciation of regional interest, it might be advantageous for us to consider a more equitable system of exhibiting the breed nationally."

At the December, 1966, meetings of the Extension Committee and the full Association Board, decisions regarding the national show locations were made. At the Extension Committee meeting, representatives and delegations from eight state fairs or expositions interested in being designated as a "National Show" were present. All had merit and the Board's final choices were based almost entirely on location.

The Western region had previously held a conference and submitted a report confirming full agreement—Colorado abstaining from the vote—on Sacramento, CA, as the site for the 1967 Western National. Other locations receiving the approval of the board were the Pennsylvania All-American Show at Harrisburg for the Eastern National and the World Food and Agricultural Exposition at Madison, WI, for the Central National. The latter two locations have remained

The National Holstein Futurity for 3-year-olds, held at the Ohio State Fair in Columbus, is now in its 31st year. Premiums paid range to over $1000. It attracts entries from coast to coast and has 1236 heifer calves entered for 1987.

for the Eastern and Central Nationals ever since. The Western National has been in varied locations until recently; it appears to have found a permanent home at Fresno, CA.

The three national show concept was inaugurated in 1967. While there was some initial opposition, over time this has faded. All the dairy breeds now have their own system of multiple national shows.

The three national show program had a positive impact on the western and eastern scenes. Many western breeders and potential exhibitors had felt their shows did not carry the weight accorded to the midwestern and eastern events. "Serious" showmen were few, and fewer still were those who made the long and time-consuming trek to the National Show at Waterloo, IA.

Now, the designation of a western show as "National" with the resulting publicity and prestige, has effected a dramatic change in the western outlook. Breeders accepted the challenge of staging a national event and actively encouraged interstate competition. The ever-increasing activity in all phases of the registered Holstein industry in the West, particularly California, over the past dozen or more years is likely due, in some part, to the start of a National Western Show.

In the years immediately prior to the Eastern National, Pennsylvania breeders, as a group, were somewhat parochial with respect to their showing. Interest was not high, and only on occasion was there a Pennsylvania State herd shown at Waterloo or at the Chicago International. Once the Eastern National Show was held in Harrisburg, PA, a resurgence in show activity took place which continues unabated.

Production Recognition

No sooner had the three national show concept begun, when an idea called a "Performance Qualified Show" program narrowly won approval at the December, 1970, meeting of the Holstein Association's Board of Directors. Intended to apply principally to the three national Holstein shows, the "Performance Qualified Show" program as adopted specified, in part: "All females in milk (excluding 2-year-olds) to be accepted in the show must have one record equal to or exceeding a production level of breed average for milk on a DHIR or DHIA 2X 305-day M.E. (mature equivalent) basis..." Females of younger ages were to qualify on their dams' records at the level stated. However, bulls' dams' records were to be 1000 lbs. of milk above the level for the females.

The Board's report had other information including, "These requirements are to be voluntarily initiated by each show...preferably within four years."

At the 86th annual meeting of Holstein-Friesian Association of America held on June, 1971, in Des Moines, IA, it was apparent that the majority of the delegates were against any production requirements for participation in the show ring. At that meeting, a resolution sponsored by the New York delegation was presented. It said, in effect, that "Production Qualified" (PQ) shows should be "opposed" by the membership although recognition should be given for high-producing show animals.

Chief proponents of the New York resolution opposing "Production Qualified" shows (somewhere along the line the word "production" had supplanted the Board's term of "performance") were Tom Coyne, R. Peter Heffering, and Avery Stafford, all of New York. Favoring the inclusion of production requirements for the shows were Tom Sawyer and Tom A. Nunes, Jr., both of California.

When the New York resolution was brought to vote, it passed with 135 for and 53 against.

The 1971 Western National Show was held in October at Fresno, CA. True to the principles that had been endorsed by two of the California delegates (Sawyer and Nunes) performance qualifications were mandatory. In the long history of the show ring, this was the only major show in the U.S. to ever take this approach. The show was high quality but lower than normal in the number of animals shown, as some people withheld animals on principle.

The efforts of the proponents of "Performance Qualified" shows were not entirely in vain. While opposed to mandàtory production requirements for shows, many breeders and exhibitors felt that the superior producing abilities of Holsteins should be spotlighted for advertising and promotional purposes. So, at the 1971 Central National Show in Madison, WI, the higher producing cows of milking age in each class were recognized.

Following the 1971 show season, the Holstein Association Board took another look at the question of production recognition in the show ring. The Board removed its approval of "Performance (or Production) Qualified" shows, substituting a format for "Performance (or Production) Recognized" shows. The Board also recommended that the national show catalogs include official production records, DHIR or DHI, on all animals with such records. Since that time, all three nationals and several other major shows have continued the practice of production recognition. In addition, following the Board's direction, it has become usual at the major shows to list an animal's best production record along with her name, birth date, owner, and other essential information in the show booklet provided for the spectators. Frequently, the production award is given to the highest producer in the first 10 placings. The object is to recognize a production winner with suitable type traits—an animal that typifies the ultimate in a parallel breeding program.

The Pennsylvania All-American Dairy Show has stimulated new interest in Holstein showing in that key dairy state.

The Three National Shows
The Eastern National Holstein Show

The Pennsylvania All-American Dairy Show made its debut in September of 1964 at the Farm Show complex in Harrisburg. It was an instantaneous success with some 2400 dairy cattle of the major breeds (including 379 registered Holsteins in open competition) from 16 states and Canada. It was reported as "the largest international dairy show in North America, if not the world." Since 1967, Harrisburg has been designated the Eastern National Holstein Show by the Holstein Association of America.

The sponsors of the show were the PA Department of Agriculture, PA Farm Show Commission, and the PA Dairy and Allied Industries Association.

The purpose of the Pennsylvania show was outlined in the first catalog of the show over the signatures of PA Department of Agriculture Secretary Leland H. Bull and Cuthbert Nairn, Sr., first chairman of the show. To quote, in part:

"The Pennsylvania All-American Dairy Show is a new venture. It is designated to fill a need that has been apparent for some years. Until now our [Pennsylvania] dairy farmers have had to take their cattle to other states to benefit from the comparisons and competition that open national shows provide. Farmers and breeders from other states and Canada have never had the opportunity to bring their best animals into competition with native Pennsylvania stock until now..."

The late Cuthbert Nairn, Sr., noted Pennsylvania Ayrshire breeder and internationally recognized dairy judge, is credited with initiating the Pennsylvania All-American Dairy Show. For years prior to its inception, he worked diligently on the project, eventually receiving the enthusiastic support of then Pennsylvania Governor William Scranton, the PA Secretary of Agriculture, various legislators of the Commonwealth, and the PA dairy breed organizations. At the time, Nairn was president of the PA Dairy and Allied Industries Assn., which continues as a

The Central National Holstein Show has been held in conjunction with World Dairy Expo since 1967. The 1968 Grand Champion, Lake Aire Dora Crisscross, is pictured with Gene Nelson, then a National Director, and Alice in Dairyland, Bobbi Thoreson.

sponsor to date and, since 1971, with well known Holstein breeder Obie Snider as president.

With $78,000 in premiums for its six dairy breed shows (five designated as "National"), shows for junior exhibitors, the Annual Youth and Intercollegiate Dairy Cattle Judging Contests, several breed auction sales, and spectators numbering in the thousands from this country and others, the 1983 Pennsylvania All-American added further luster to its impressive record.

While many individuals are responsible for the show's success over the years, the overall managers of record have been but three — Harold R. McCulloch, 1964-68; Thomas W. Kelly, 1969-73; and, currently, Charles A. Itle since 1974.

The Central National Holstein Show

World Dairy Expo began as the World Food and Agricultural Foundation, Inc. The first show was held in 1967 at the Dane County Exposition Center in Madison, WI, its permanent site. The show ran for several years as a combination effort with big name entertainment and lost money.

In 1971, World Dairy Expo was incorporated. It was streamlined to become a trade show for dairy farm families. World Dairy Expo's schedule of events included a list of activities for dairy farmers, dairy farm wives, 4-H and FFA members and dairy enthusiasts from other countries. With activities focused on dairy cattle, the show began to prosper and gain legitimate recognition in dairy circles.

The late Allen Hetts was instrumental in getting World Dairy Expo launched. Since then, many Holstein and industry leaders have contributed their abilities to the World Dairy Expo Board of Directors. They include, but are certainly not limited to: Gene Nelson, Gray View Farms; Arthur Nesbitt, Nasco; Greg Blaska, AMPI; Robert E. Walton, ABS; M. B. Nichols, sales managers; Ted Krueger, Curtiss; Alton Block, Tri-State Breeders; R. Dale Jones, Sunnymede Holsteins; Roy Hetts, Crescent Beauty Holsteins; Tom Lyon, Midwest Breeders; Jack

The youngest national show is the Western International Dairy Exposition.

Bingham, Golden Oaks Farms; James Crowley, University of Wisconsin; Verlo DeWall, AMPI; and Ivan Strickler, Strickler Holsteins.

World Dairy Expo's series of cattle shows draw visitors from around the world. About 1600 show animals are housed annually. Five of the six breed shows, including the Central National Holstein Show, have a national show status. Only the Jersey show is a regional one. The World Premiere series of Holstein sales also contributes a good deal to the success of this show. With press estimates of 5000 in attendance, it boasts sale averages of five figures and sale toppers of six figures. Cattle have been sold to over 30 states and 10 foreign countries.

Youth, too, play an important role in the growth of World Dairy Expo. Expo hosts both the National 4-H and Intercollegiate Dairy Cattle Judging contests, some national junior breed shows, a series of FFA dairy cattle and dairy product judging contests, and the Central National Junior Showmanship contest. Additionally it hosts bus loads of Vo-Ag youth spending the day viewing the best dairy technology and cattle available.

B. D. Craig was the Executive Vice President from 1968 through the 1979 show. Brad Rugg has been Expo's Executive Vice President since 1980. From nothing more than an idea, World Dairy Expo has, in less than 20 years, emerged into a major dairy event with contacts throughout the world.

The Western National Holstein Show

Over its 18-year history, the Western National Holstein Show has been held in Sacramento, CA, with the State Fair twice (1967 and 1968), in San Francisco in conjunction with the Grand National Show in 1976, with the Western Washington Fair in Puyallup three times (1973, 1975 and 1977), and in Fresno, CA, the other years.

The Western National, regardless of location, has been popular. However, some felt that the lack of a permanent home was preventing the show from attaining true national stature, so exhibitors from western states resolved to upgrade the show to make it comparable to the other nationals (Central and Eastern). In May, 1978, then National Holstein Association Director Henry Polinder of Washington invited the western state associations to send representatives to a meeting to discuss the Western National. A show committee, chaired by Tom A. Nunes of California, decided to organize a show that would be held independent of other local and state fairs, such as was being done at Madison and Harrisburg. Doug Maddox, president, and Greg Bovre, executive secretary, of the California Holstein Association, were instrumental in establishing the Western National Show, along with many other prominent western breeders.

At its January, 1980, annual meeting, members of the California Holstein Association voted to set up what was to become the Western International Dairy Exposition (W.I.D.E.), utilizing the facilities of the Fresno County Fair at Fresno, CA. This proposal received the active support of breeders in the other states of the western bloc.

Superintendent of the first show in October of 1980 was the popular and capable Robert Selkirk, retired professor of dairy science at the University of California-Fresno, who had played a major role in the organizing of W.I.D.E.

The 1980 Western National set a new pattern for the series and was highly successful. There were 307 Holsteins from 15 states and Canada, with eight groups in the state herd class. Visitors were recorded from 31 states and nine foreign countries. To encourage wide participation, travel money was paid for each animal (based on geographic zones) and class winners at the Eastern and Central Nationals were offered up to $1000 to compete at Fresno.

The unexpected death of Professor Selkirk in August, 1981, was a blow to the W.I.D.E. organization but the show has contracted out the management for the last several years. The Western International Dairy Exposition has proven to be a success during its short history. The states directly involved in its operation are working to get the organization on a sound financial base with the expectation that W.I.D.E. will remain a positive force during the years to come.

The aged bull class at the 1908 New York State Fair.

State Association Shows

Competition at national Holstein shows is beyond the reach, financially and time-wise, for most farmer/breeders. The top animals are so closely matched that often a win depends on showmanship, the fitting of the animals, as much as the animal itself. This has been true for decades, and has resulted in a two-tier system, similar to public auctions.

In the last three decades, a network of local, state, and regional shows has developed and flourished to the point where it has become the backbone of the Holstein show program. Here, a breeder can show milking animals, keeping them away from home only a day or two, perhaps, and come home with a ribbon or better yet, a solid offer for one of the animals. Also, there is always the chance of having an animal "discovered" by someone who can take it on to fame and glory. More than one All-American has been first found in the local show ring.

The state program started 70 years ago. In March, 1915, the Holstein breeders of the Cache Valley in Utah held a one-day show of their Holsteins in Richmond, which they called a "Black and White" Show. Little did those early Utah breeders realize that their pioneering effort would lead to today's far reaching program of state association sponsored shows, completely separate from the state fair shows.

Richmond, UT, was the site of the first state association show in 1915. Shown here is the aged cow class at the 1951 show.

In 1915, the Richmond Black and White Show was the only one of its kind. In 1983, there were 41 state association sponsored Holstein shows held in 38 states. In these state shows, nearly 9000 individual animals were exhibited (this number does not include the three national shows, nor any state fairs or expositions).

Today's sophisticated shows with, in many cases, a multitude of classes and production recognition for milking animals are a far cry from that first Richmond, Utah, show in 1915.

In the April 27, 1940, *Holstein-Friesian World,* Prof. George B. Caine, head of the dairy department at what was then the Utah State Agricultural College (now Utah State University), described that first Richmond show:

"The section around Richmond was the center of Utah's breeding establishments and much was done to advertise the black and white cattle. In order to do more advertising, a group of leading breeders decided to hold a spring show patterned after the European parish shows...the date was set to March, 1915.

"At that time, cattle were not led and handled as they are today. They were driven in large groups by men on horseback. It took some time to decide how to stage that first show. At first it was thought all the cattle — perhaps as many as 500 head — should be driven up and down the streets of the city so all the people could see what was being developed. Finally, it was decided to bring the cattle in small groups by owners and put them in yards that were built inside a large enclosure owned by the church.

"The judges went from one yard to another, pointing out the better individuals. No classes were followed in a regular judging program. The bulls were led individually and walked about in a general way, but not officially placed.

"There were few young animals, but mostly cows in milk. There were some 250 head, not selected nor fitted, but driven from the barns at home to the central location, so there were more numbers than quality."

From this beginning came today's nationwide program of state association sponsored Holstein shows which enjoy such tremendous popularity. Yet, the concept of the Richmond Black and White Show was slow to catch on among other states as the big state fair shows continued to flourish. Some 20 years after

The 1953 Michigan state association Black & White Show.

the Richmond Show's founding, no other state shows were reported, although the *Holstein-Friesian World* magazines of the time carried numerous reports of county shows. Also, a few statewide shows were occasionally held in conjunction with summer picnics or field days. Other shows, such as the Southern Minnesota, Northern Wisconsin and Eastern Indiana shows surfaced in the mid-1930's.

In 1937, Illinois held a state show and picnic, and Iowa had a state show limited to the champions of the nine district shows held earlier in the season. The Iowa Show was repeated the following year.

After this leisurely start, the pace quickened, and by 1969, there were 35 high caliber state association sponsored shows, growing to 41 such shows in 1983. In several states, the state event follows a series of regional or county shows in which the cattle receive their first screening. The result is high quality in the state show, and even higher in the state fair which may follow.

The Role Of The Holstein Association

Prior to establishing the three national shows, the Holstein Association Board of Directors did not take an active role in the show program.

With the abrupt termination of the National Show at Waterloo, the Association officers and Board immediately saw the need for action and changed from a "passive" to an "active" role in Holstein showing. The Board's efforts have been directed primarily to the three officially designated national Holstein shows. A few shows are officially designated as "regional" by the board, with the intent of providing additional prestige and thus encouraging greater participation. The Pan-American (State Fair of Texas at Dallas) has been the Southwestern Regional, while Southeastern Regional has also been named, on a rotating basis, in the Carolinas, Tennessee or Kentucky.

Policies set by the board for the nationals have included formulating the minimum number of classes required; production recognition for the animals shown; uniform birth date requirements for the classes; writing and implementing an updated Show Ring Code of Ethics; aid to show management in procuring competent judges (but not the actual selection); and the naming of one person to serve as the "associate judge" at all three nationals annually.

In recent years, the Holstein Association board has initiated judges' conferences for the exchange of ideas to promote uniformity in evaluation and ring procedure.

These efforts have paid off. The regulations and practices set by the national shows have set the pattern for most of the other shows in the U.S., resulting in quite uniform regulations and procedures.

The Holstein Association's involvement in the show program is not limited to policies or "official" designations. Substantial financial and personnel support is also provided, amounting to some $39,000 for 1983.

This includes payment of the national show judges and their expenses, money for state herds in the national shows (varying amounts based on mileage from the state of origin), and $100 for each state herd shown in a regional show. Association staff, as many as 10 to each national, are assigned specific duties to aid in the conduct of the shows.

As in every other aspect of the Holstein industry over the past quarter of a century, the Association has been in the forefront in an effort to improve the entire concept of showing.

The All-American Program

For 60 years, the annual selection of the All-American winners has been the climax of each show season. Designation of an animal or group as the All-American of the year is a goal offering the highest prestige available in connection with the show ring, providing worldwide recognition for the animals, their breeders and/or owners. Further, for one interested in merchandising, it is proved year after year that the title of "All-American" opens the door to an almost unlimited market, both domestic and overseas, at premium prices. But, regardless of the enhanced marketing opportunities, to many the fact that he or she has bred or owned an All-American is sufficient reward in itself.

The All-American program, sponsored throughout its history by the *Holstein World,* was born in 1922. Following the 1921 show season, the *World* featured a Show-Cow-of-the-Year selection. This simple and unofficial effort aroused so much interest that it was decided to try to designate the top animal or group for

Triune Papoose Piebe made All-American six times, starting as a calf in 1926.

the nation in each class by comparing the classes in the major shows of the year, following the same idea as the All-American football teams chosen in those days by Walter Camp.

That first year, 1922, the selections were made by W. S. Moscrip, noted Minnesota breeder and judge who had placed the classes at the 1922 National Show, the Pacific International, the Eastern States Exposition and Wisconsin State Fair, among others. In making the All-American selections, he was assisted by Maurice S. Prescott, editor of *Holstein-Friesian World* and the originator of the All-American idea for livestock. The selections covered each individual class plus the get of sire and produce of dam groups.

The idea of postseason awards proved so popular with the breeders it was decided to make the selections on a broader basis beginning in 1923. Instead of having a single judge, selections were made by a committee consisting of the judges for the year at the National Show, Waterloo, the Pacific International (at Portland, OR) and Eastern States Exposition (W. Springfield, MA), with the editor of the *World* an ex-officio member. The committee met at the end of each season and made the selections in personal conference.

This plan worked reasonably well, but with the growth of interest and participation in the show ring, and particularly the increase in numbers at the major shows, it became obvious that more judges of shows around the country should be involved in the selection process.

Accordingly, in 1939 the present system of selection was initiated. Exhibitors submit the pictures of their winning animals or groups at state, regional and national shows to the *World.* A nominating committee of three judges (in recent years, the judges of the three nationals) then meets and from the pictures, supplemented by their personal knowledge of the entries, sort out what they consider to be the six best in each class, depending upon the judges' opinions as to the strength and depth of quality in the class involved. Nomination in itself is a signal honor, regardless of the outcome of the vote of the entire All-American panel.

The pictures and all show placings of the nominated entries are then reproduced in booklet form and mailed to the panel of All-American judges. This

Harvey Swartz has nominated All-Americans for nearly 30 years.

panel includes 25 or more of the judges of the leading shows of the year. They vote independently by mail, indicating their first, second and third choice in each class. Point values are assigned as follows: Each first place vote, 7 points; second place, 3 points; third place, 1 point. By simple arithmetic, the awards for All-American and Reserve All-American are determined in order of total points. Every other nominee receiving 16 points or more is named Honorable Mention. Third place winners are automatically High Honorable Mention.

While the results represent a composite vote, the judges making up the panel have officiated at the strongest shows in the country. The impartiality of this method has maintained the integrity of the program.

All nominations receive extensive coverage in the *Holstein World* (as well as other publications) and even more publicity goes to the eventual winners.

This success is evidenced by the adoption of similar programs patterned after the *World* All-Americans by the other major dairy breeds—Guernsey, Jersey, Ayrshire and Brown Swiss—in the 1950's. The *Holstein Journal of Canada,* with whom the *World* has always enjoyed close relations, was the first to pick up the idea and their All-Canadians have been going strong since 1943.

For a number of years, the two leading Japanese livestock magazines, *Holstein* and *Dairyman,* have annually requested and published the pictures of the All-American winners, indicating the intense interest of Japanese Holstein breeders in U.S. Holsteins.

As expected for a program of some 60 years duration, there have been changes from time to time. For example, classes designated for All-Americans have followed the general rules of the shows themselves.

With the advent of widespread use of artificial insemination (AI), private ownership of herd sires declined and the shows found fewer bulls in competition. Thus, in 1951 the aged bull class was changed to include bulls 3 years of age and older, eliminating the separate class for 3-year-olds. And, for 1983, the nominating committee, after considering the field, made one class for bulls

2-years-old and over. At this writing, due to changes in the bull classes at some of the shows, the future of classes for bulls over the age of senior yearlings — and perhaps for all bulls — is in question.

Originally, only winners of their classes in the major shows were considered. However, with the continued stiffening of show competition, and especially with the three national show program in effect, it was felt changes should be made to ensure eligibility of all worthy animals. Currently, eligibility of animals and groups is determined as follows: (1) a placing of first through fifth at a national show; (2) first and second place winners at any U.S. show in which 300 or more Holsteins are in competition, or at which not less than six state or provincial herds are shown; and (3) first place winners at state fairs, regional shows and state Holstein shows at which less than 300 head are in competition.

With the high interest expressed in the All-American competition, it was logical to establish a "Popular Judging Contest." Conducted by the *Holstein World* and with substantial prizes for the winners, entrants send in their own selections for All-American and Reserve honors after studying the photos and show information on the nominations published in the *World.* The winners are those whose entries most closely correspond to the judges' panel vote. This has been a popular annual feature since the first such contest in 1937.

An All-American Update

Triune Papoose Piebe is the only female to be named All-American in every class, and she made it six times starting as a calf in 1926. Bred by Hargrove & Arnold, Norwalk, IA, she was exhibited by them in 1926 and 1927, before her sale to Mount Victoria Farms, Hudson Heights, Quebec. They later showed her to All-American at 2, 3, and 4 years and as an aged cow, 1928 through 1931.

The only male to duplicate this feat — winning All-American in every age class — was King Bessie Senator, bred by Robert V. Rasmussen of Elmwood Farms, then with the Deerfield, IL, address. Elmwood showed the bull as a calf in 1937 and as a senior yearling, 1938. From 1939 through 1942, King Bessie Senator was exhibited by Ravenglen Farms, Antioch IL, and he won as a 2-year-old, 3-year-old and twice as an aged bull. The sire of King Bessie Senator was King Bessie

The show ring is where All-Americans first earn their laurels.

Ormsby Pietertje who has 13 sons and daughters accounting for 25 All-American awards, not including succeeding generations.

In addition to his own two All-American Get of Sire groups (1947 and '48), Wisconsin Admiral Burke Lad's sons and grandsons dominated the Get of Sire winnings during the late 1940's and into the mid-fifties. For three years in a row, 1949-50-51, Pabst Farms, Oconomowoc, WI, won both the All-American and Reserve Gets. Pabst Regal and Pabst Roamer, both Burke Lad sons, won in 1949; Roamer and his sire in 1950; Regal and Roamer again in 1951. In 1952, Roamer's get was All-American; in 1953 the Get of Pabst Comet (Burke Lad son) was Reserve; in 1954 the Get of Weber Hazelwood Burke Raven (Burke Lad grandson) was Reserve; in 1955 the Get of Wis Leader (another Burke Lad grandson) was Reserve and in 1956 the Leader Get was All-American. In all, Pabst Farms showed eight All-American Gets, including the Get of Carnation Sensation in 1938.

During the 1930's, Maytag Dairy Farms, Newton, IA, was another power in the show ring. Their gets of Man-O-War were four times All American from 1931 through 1934, and, not consecutively, they showed nine Produce of Dam groups to All-American honors.

The building blocks for the brightest chapters in 25 years of recent All-American history are quite simply ABC's. Perhaps no other sire has written a loftier legacy for the breed than ABC Reflection Sovereign (EX-Extra in Canada). In 1954, he became the first sire to have three milking All-American daughters: ABC Shamrock Harriet R (2-yr-old, 1951), ABC Shamrock Mildred (4-yr-old, '53, and aged cow, '54), and Rosafe Shamrock Kit (4-yr-old, '54). And, in 1962, he became the first to sire five All-American Gets. These uniformly excellent quartets won top honors in 1953-54-55-57-62, with the latter group selected Reserve All-Time All-American, 1984. Others of his gets were Reserve in 1960 and '61, plus Honorable Mention, 1959. In all, his offspring collected seven individual All-American awards.

While the daughters of ABC Reflection Sovereign shone brightly on the tanbark, it was his sons that excelled in siring even more winners. Heading the list by a phenomenal margin is Romandale Reflection Marquis (EX-95-SMT) who, starting in 1966, accounts for 29 individual All-American designations on 20 sons and daughters, as well as 14 Reserves. His best year was in 1968 when he sired seven of the 12 individual winners with an ABC grandson, Thornlea Texal Supreme, accounting for three more.

With the exceptionally wide AI use Marquis received, it was inevitable that his sons would continue his heritage of siring All-American type. Heading this department with seven awards for his offspring is Agro Acres Marquis Ned, followed closely by Prestige of Lakehurst with six. With one each are Fultonway Randall, Jergens Distinction Marquis, Puget Sound Highmark, Whippoorwill Marquis King, Goma Marquis Jay and Agro Acres Unique. Another prepotent ABC son was Rosafe Citation R with three sons siring a total of seven All-American winners.

If records are made to be broken, it is only fitting that Ideal Fury Reflector (EX-94-GM) should be the one to surpass his own grandsire.

In 1973, Fury made his mark in the all-time record book when Gene Acres Felicia May Fury, pride of the late Allen Hetts' Crescent Beauty Farm, became his fourth milking daughter to be named All-American. Preceding Felicia May were Easthaven Quintillion Fury (2-yr-old, 1970), Rosemere Fury Iva (4-yr-old, 1972) and Polytechnic Fury Ronda (3-yr-old, 1972). For good measure, Fury also sired the Reserve All-American Aged Cow behind Felicia May in both 1973 and 1974, plus the High Honorable Mention winner in 1973, an accomplishment shared only by Bond Haven Rag Apple Maple in 1959 and Osborndale Ivanhoe in 1969. In all, nine Fury sons and daughters claim 11 All-American awards. But, it was his gets in the early 1970's that took him to even greater heights, as they were in the top spot for five consecutive years with the 1972 group selected as All-Time All-American Get of Sire in 1984.

The early 1970's may best be remembered for the fast paced and electrifying clashes of two titans of the tanbark when Zeldenrust Fond Memory (EX-97) and C Carlspride Vogel Reflection (EX-97) met at center stage at the Central National Show. Fond Memory, a Marquis son shown mainly under the Cash-Mar banner of C. M. Bottema, Jr., of Indiana, and Carlspride, a Citation R grandson brought out by Hi-Path (Henry P. Bartel), Pinehurst (David Bachmann), both of

Zeldenrust Fond Memory and C Carlspride Vogel Reflection met in the ring many times during the early 1970's.

Gray View BD Crissy was unmatched as a brood cow in the 1960's.

Wisconsin, and James A. Walker & Sons of Ontario, wrote some interesting history themselves. When Fond Memory's career came to a close in 1976, he had been nominated for All-American nine consecutive years, amassing five All-American awards (four as an aged bull and, in 1968, as a 2-yr-old) as well as two Reserves and two Honorable Mentions. Just for good measure, he was twice All-Canadian and three times Reserve. His long-time opponent, Carlspride, finished up his six years on the circuit with three All-American awards, two Reserves and an Honorable Mention.

If ABC Reflection Sovereign is the king of the All-American parade, then Gray View BD Crissy reigns supreme as the queen. Bred by Harvey A. Nelson & Sons, Union Grove, WI, this Excellent-93 Gold Medal Dam by Papst Burke Tritomia Don has records to 867 lbs. fat and was Reserve All-American 2-Yr-Old in 1952. But, it is her unparalleled brood cow status through the 1960's that brought her such fame.

In 1966, she became only the second cow in history to be the dam of three All-American winners. (May Walker Ollie Homestead is the other, receiving her honors in the mid-1930's.) Crissy's honored progeny were Gray View Pet Crystal (sr. yrlg. heifer, 1955; 2-yr-old, 1956; plus reserve 4-yr-old, 1958), Gray View Crisscross (aged bull, 1962, 63, 64, 65 and res. 2-yr-old, 1961), and Gray View Pet Crysta (4-yr-old, 1966).

While the Crissy distaff side was basking in glory, her son, Gray View Crisscross (EX-96-GM), by Bond Haven Rag Apple Maple, was cutting a wide swath in the ring, both on his own and through his offspring. The All-Time All-American Aged Bull winner in 1966, he still found time to sire winners of seven All-American and eight Reserve awards along with the top vote-getting gets in both 1967 and '68. He repeated as All-Time All-American in 1984.

Of all the All-American classes, the aged cows were always considered to be the "creme de la creme." In the 59-year history of the competition, only three herds have brought out three aged cow winners. These premier exhibitors were Mount Victoria of Canada with Oakhurst Colantha Abbekerk, Triune Papoose Piebe and Montvic Rag Apple Bonheur (twice), all three of whom left lasting images on the breed; Paclamar Farms of Colorado with Harborcrest Rose Milly (three times and whose son, Paclamar Astronaut, has seven All-American offspring), Green Banks Admiral Mooie and Kanza Matt Tippy; and R. Peter Heffering in partnership with James J. Houlahan of Tara Hills in New York with Maroy Model Abbekerk and Johns Lucky Barb, and in partnership with Kenneth Trevena of Hanover Hill (first of New York, now Canada) with JPG Standout Kandy (three times).

Six cows have been selected as All-Americans in the climactic aged cow class three times. They are Rosehill Fayne Wayne for J. Durno Innes, Canada in 1946 and Franlo Farms, Minnesota, 1948-49; Rocky Hill Mont Burke Dusty Jo for Dr. Harold J. Schmidt, Lavacre, CA, in 1950-51-52; Linden Dictator Wimble Wimpy for Braun's Sunny Lea, Wisconsin, in 1958 and Nelson Rehder (WI) in 1960-61; Harborcrest Rose Milly for Paclamar Farms in 1962-64-65 and Reserve All-Time; Northcroft Ella Elevation for Woodbine Farms, Pennsylvania, in 1980 and for Woodbine and Romandale Farms Ltd. of Canada in 1981-82; and, as mentioned in the paragraph above, R. Peter Heffering's JPG Standout Kandy, 1977-78-83 and Honorable Mention All-Time.

All-Time All-Americans

There have been three All-Time All-American contests. The first was during World War II, in 1943, when wartime restrictions caused the cancellation of the show season. To offer show ring "addicts" their "fix", the *World* conducted the first All-Time All-American contest covering 20 years of All-Americans...1922-1942.

Again in 1966, a new round of All-Time selections were made, after another 20 years of showing. Every previous All-Time winner was deposed.

In conjunction with the centennial observance of the Holstein Association, the *Holstein World* conducted an All-Time contest in late 1984. The awards were confined to the seven more important classes: the milking females, produce of dam, get of sire, and aged bulls. A two-step process was followed just as in the All-American contest.

Eleven judges, who had served on the All-American panel 10 times or more during the 1966-1983 period, were joined by three veteran judges. This panel made the nominations and then pictures of the nominees were mailed to 73 judges who voted.

All-Time All-American Nominating Committee

Richard Keene	18 years
Fred Foreman	17
Jim Lewis	17
John W. McKitrick	17
John Morris	15
Henry "Sonny" Bartel	14
Donald V. Seipt	13
Ray Brubacher	11
Gene Nelson	11
Charles Norton	11
Tom Nunes	10

Plus three veterans:
Harvey Swartz (8 in this period and 27 in all, the most of anyone)
A. C. "Whitie" Thomson
George Trimberger
(All three served on the 1966 All-Time panel, as did Nelson and Norton)

Veteran judge, Richard Keene, has picked All-American nominations for two decades.

Considering the exciting show ring contenders in the 1960's and 1970's, it came as no surprise that all but one previous All-Timer was dethroned, supplanted by animals of the current era.

Northcroft Ella Elevation (3E-97), All-Time All-American Aged Cow, is considered by most to represent the ultimate in U.S. Holsteins.

Brookview Tony Charity (EX-97), All-Time All-American 4-Year-Old.

Aged Cows

Northcroft Ella Elevation (3E-97) represents the ultimate in a breeding program stressing both production and type. One of only four animals to win All-American honors four times in milking form, she won the All-Time All-American Aged Cow class handily. She is the highest record All-American with top credits of 7-7 364d 48,731 4.2% 2028. She triumphed at World Dairy Expo in 1980 when she bested the Holstein entries and was named Supreme Champion over all breeds. Also earning the Reserve designation in the All-Time 3-Year-Old class, she is the exhibit of Woodbine Farms (the George Knight family) and Romandale Farms Ltd., Airville, PA, and Unionville, Ont.

Voted a solid Reserve is Harborcrest Rose Milly, also a high producer with 1242 lbs. fat, and a Gold Medal Dam. She had easily won the All-Time Aged Cow in 1966 and proved popular with the judges this time, earning a nod from 67, the most of any animal in the contest.

Finishing third in the voting, good enough for High Honorable Mention, was Gene Acres Felicia May Fury (5E-97), sired by Ideal Fury Reflector and a member of his All-Time All-American Get of Sire.

It is truly a fabulous class featuring the big hitters of a glamourous era in the show ring. And they produce, too, with the 12 nominated aged cows' best records averaging 29,956 4.0% 1204.

All-Time All-American 3-Year-Old is Shadowcliff R A Gina (EX-97).

4-Year-Old Cows

Brookview Tony Charity (EX-97), a current star of the tanbark in 1985, made it an easy win in the 4-year-old class with 338 points, the most of any class winner. She was All-American and All-Canadian as a 4-year-old in 1982. Exhibited as a 4-year-old by George Morgan and Hanover Hill Holsteins, Walton, NY, and Port Perry, Ont., she has a top record of 4-11 343d 3X 37,340 3.5% 1310.

In Reserve is Shadowcliff R A Gina, the popular Elevation daughter who won All-Time designation in both the 3-year-old and 2-year-old classes. More on her below.

3-Year-Old Cows

Shadowcliff R A Gina (EX-97) topped this strong class, credited to her sire, Round Oak Rag Apple Elevation, and exhibitor, Bower Farms, Inc., LaGrangeville, NY. She compiled a most successful record in All-American competition, winning the 2-year-old, 3-year-old, and 4-year-old classes in successive years in 1979, 1980 and 1981. With only two points separating Reserve and High Honorable Mention, Northcroft Ella Elevation edged the previous All-Timer, Pat Willow Lake Victor.

2-Year-Old Heifers

As noted, the Gina heifer was named the All-Time winner here. Also as noted above, she went on to bigger and better things after her 2-year-old season when she was brought out by Catskill Registered Holsteins. The heifer named Reserve All-Time 2-Year-Old also had a remarkable career. Wapa Bootmaker Mandy went on to produce 4-3 365d 41,420 3.3% 1378, the second highest milk record of any All-American. She classified 3E-96 and earned her Gold Medal Dam rating, as well. This Paclamar Bootmaker daughter was exhibited by Alison Place and Phil Fisher, Wapakoneta and Fort Jennings, OH.

Aged Bulls

By including the aged bull class in 1984 All-Time competition, the traditional

The All-Time All-American Produce of Dam is C Glenridge Citation Roxy and C Glenridge Emperor Rocket.

importance of the class over the last 20 years was reflected. But the fact that the 1984 competition was won by the previous All-Time winner from 1966, showed the changing attitudes towards exhibiting older bulls. Gray View Crisscross was a convincing winner here for Harvey A. Nelson & Sons, Union Grove, WI. Four times a winner of All-American classes in the early 1960's, he is one of the offspring of his dam, Gray View B D Crissy (EX-93-GMD), that have collected a total of seven individual All-American awards.

In the contest for Reserve, Zeldenrust Fond Memory edged his long-time show ring nemesis, C Carlspride Vogel Reflection. Fond Memory, shown by the late C. M. "Cash" Bottema, of Indianapolis, IN, was All-American four times in 1970, 1974, 1975, and 1976. Carlspride, exhibited by Lakehurst and James A. Walker & Sons, Sheboygan Falls, WI, and Grimsby, Ont., was All-American in 1971, 1972, and 1973.

Produce Of Dam

A pair of great cows, high-scoring high producers, earned All-Time designation in a fairly tight fit with the previous All-Time Produce. These daughters of C Norton Court Model Vee (EX-6*) were All-American Produce three consecutive years. C Glenridge Citation Roxy (4E-97-GMD) with 1166 lbs. fat has been a favorite of Holstein enthusiasts everywhere. Her sister, C Glenridge Emperor Rocket (3E-96) with 835 fat, has likewise, been popular at shows. Bred by Lorne Loveridge & Family, Grenfell, Sask., they were jointly exhibited with Robert & Craig Miller, Mil-R-Mor Farm, Dundee, IL.

Get Of Sire

Just five points separated the All-Time winner and Reserve in this class, the closest point spread in the contest. The get of Ideal Fury Reflector, All-American for five consecutive years in 1971-75, earns All-Time status.

The previous All-Time winners, the get of ABC Reflection Sovereign, are Reserve All-Time. The gets of this great sire were All-American five times as well and his daughters rewrote breed history. Fury, of course, is his grandson.

The Get of Ideal Fury Reflector is the All-Time All-American winner.

The All-Time All-American selections of 1984 provide a fitting capstone to the shows of the last 100 years. The graceful animals put the old adage, "show cows don't milk," to rest with their sparkling production records.

It has been 62 years since Maurice Prescott came up with the idea of an All-American program. Many aspects of the Holstein industry have changed drastically since then but the All-American program remains as popular as ever. Every year, as soon as the issue of the *Holstein World* containing the nominations hits the mailbox, it is pored over by novice and veteran Holstein enthusiasts alike. 4-H youngsters, prominent breeders, as well as Holstein Association executives, all make their own All-American selections in the annual "Popular Judging Contest." In an age of embryo-splitting and on-farm computers, the All-American program is a tradition of the registered Holstein industry that has stood the test of time.

Accomplishments Of The Show Ring

Throughout the history of the Holstein breed in the United States, the issue of production versus type has generated many a discussion among breeders. Looking at 100 years of breed development, the pendulum has swung from one extreme to the other. "Production is basic and must receive first consideration; however *practical* type must receive proper attention in order to capitalize on the high production bred into the modern dairy cow," wrote George Trimberger, the noted cattle judge and professor. One of the chief accomplishments of the show ring of the past century has been to focus attention on type traits.

As noted earlier, breeders were enchanted with production records after the turn of the century and disregarded type, often while scoffing at show ring devotees. Those attitudes changed with the Cabana debacle. A similar situation occurred after the high producing years of World War II. While herd classification was now helping to establish type, breeders were again concentrating on production. The show ring would, under the direction of a number of cattle judges, swing the Holstein breed toward dairy character and lean, angular animals that have spread across the world.

Professor Trimberger explained the situation during a 1984 interview. "For a

George W. Trimberger, designer of Descriptive Classification, stressed the need to balance production and type. As a judge, he witnessed a shift to leaner Holsteins that was prompted by the show ring.

number of years, type was a controversial issue and for a while, it looked like a losing battle. Breeders in the 1950's were fattening their show cattle which was harmful to production, udder quality, and longevity. Then a trend started in the show ring toward the practical with an emphasis on dairy quality."

The judges wanted change and breeders got the word fast. Trimberger said, "The year [1955] that I judged the National at Waterloo, there was a wave of remarks across the country—'Bring them sharp this year and leave your fat ones at home. He likes them sharp.' Other judges were in the same mood."

That shift toward leaner Holsteins was an important contribution by the show ring to the breed. It also evolved, in the next decade, into the adoption of the Descriptive Classification program (Chapter 4) which would help make the arguments about type versus production pretty much a thing of the past.

The show ring has made many other contributions to the development of the Holstein breed. It was one of the early tools used to develop overseas markets for U.S. Holsteins as judges traveled extensively to officiate at shows in foreign lands. Over the years, countless breeders have been drawn into the Holstein ranks by shows. There is no way to know how many of today's breeders of registered Holsteins remained in the industry because of their early experience at club sponsored shows. Likewise, how many herds resulted from the purchase of the first registered calf for 4-H; how many family partnerships? The numbers are significant. The promotion and support of Holstein shows over the years by local, state, and national Holstein groups have contributed much to the advancement of the U.S. Holstein breed in the last century.

Are shows still worthwhile in 1985? One young breeder explained why she

To many breeders, merchandising, comradeship, and excitement are more important than ribbons as show ring rewards.

The show ring has started many a youth on a lifetime of breeding U.S. Holsteins.

shows her cattle: "It's fun. It's a vacation for a few days. Even if we don't place, by having a sharp exhibit, we get people looking at our cows. Maybe they'll get interested in our breeding program. It's how my folks got their start in merchandising Holsteins."

Merchandising, camaraderie, a few days off — just a few of the reasons why the show ring still draws exhibitors and spectators. Like the public auction, it is a tradition of the Holstein industry that has remained virtually unchanged for 100 years. While the 1985 animals look vastly different from their forebears, and their milk production has tripled, they still meet in spirited competition under the eye of an expert, with many dollars in merchandising value often awaiting the decision. It has been that way since the beginning.

Organizations that helped Build the Breed

Over the past century, many groups have contributed to the advancement of the Holstein breed in the United States. Some, like the Dairy Herd Improvement Association (DHIA), the National Association of Animal Breeders (NAAB), and the Holstein-Friesian Association of Canada, are well-described in other publications. The organizations covered in this chapter are those that have special ties to the Holstein Association and the U.S. Holstein industry.

The Breed Publications

A year after the establishment of the Holstein-Friesian Association of America in 1885, E. P. Beauchamp, Terre Haute, IN, began publishing the first magazine of the breed, the *Holstein-Friesian Register*. It was a yellow-covered, digest-size semimonthly that performed a worthwhile service for nearly 42 years.

The *Register* was purchased in 1888 by Frederick L. Houghton, Brattleboro, VT, and continued in his ownership until his death in December, 1927. Houghton was elected Secretary of the Holstein-Friesian Association of America in 1894, and he used the magazine as a vehicle to unseat incumbent Secretary Thomas B. Wales (Chapter 2). After Mr. Houghton's death the *Register* was purchased from his estate in 1928 by *Holstein-Friesian World*. Since that time, the *World* has been the only national publication of the Holstein breed.

The *Holstein-Frieisan World* began in January 1, 1904 as a semimonthly. Founders were Eugene M. Hastings of Lacona, NY, and Charles G. Brown, Ithaca, NY. It soon became obvious that the infant magazine would not support two families and Mr. Hastings withdrew, leaving Mr. Brown to go on alone. In

The Holstein-Friesian World *booth at the 1914 National Dairy Show. From the left, Ira S. Brown, E. M. Hastings, C. G. Brown, and M. S. Prescott. The Browns produced the Western Edition of the World; Hastings and Prescott the Eastern Edition.*

1912 Ira S. Brown, son of the founder, was put in charge of a Western Edition of the *World* at Madison, WI. This came out in alternate weeks and both editions went to the same mailing list, making the *World,* in effect, a weekly magazine, which it continued to be until 1932.

In December 1913, E. M. Hastings and Maurice S. Prescott purchased half interest in the *Holstein-Friesian World.* They produced the Eastern Edition, while Mr. Brown, Sr. joined his son to work on the Western Edition.

This partnership was terminated in the late summer of 1915, Mr. Brown exercising an option to repurchase. He had in the meantime negotiated a sale of the whole magazine to Ward W. and Ralph J. Stevens, prominent Holstein breeders at Liverpool, NY. They promptly put it in a corporate form of business organization and employed Frank T. Price as Editor and Business Manager. The Western Edition was moved to Waterloo, IA, with H. E. Colby, a former editor of *Kimball's Dairy Farmer,* as Western Editor.

Meanwhile, E. M. Hastings and M. S. Prescott launched the *Black & White Record,* "The Every Week Journal of Holstein Progress," on January 1, 1916, and it proved to be a lively entrant in the Holstein publication field. William A. Prescott, then a senior student at Cornell University, was a contributing editor.

With two aggressive weekly publications plus the semimonthly *Holstein-Friesian Register,* which operated at a less active tempo, the Holstein industry was covered in depth. But, even though business was booming under the war time economy, breeders would not support three breed journals.

Therefore, in the summer of 1918 the *Record* was merged with the *World.* M. S. Prescott came back as Editor and Frank Price remained on the *World* staff as Business Manager, both taking a financial interest in the company. Mr. Hastings withdrew from the enterprise to devote his time to his own sale and pedigree business. Paul B. Misner, who would go on to distinction at Dunloggin, had been working with Price and continued on as associate editor.

The printing of both editions was moved to Sandy Creek, NY, where The Corse Press again became the printer. The *World* slogan, "The Newspaper of the Breed," was first announced in the issue of December 21, 1918, and carried on the masthead until the name was shortened to *Holstein World* in January 1980.

Maurice S. Prescott was the dean of dairy breed journal editors. Recognized by the Holstein-Friesian Association of Canada with its first Certificate of Superior Accomplishment in 1956, he also received similar awards from the Holstein-Friesian Association and state and local Holstein clubs. He served on committees for the national and New York Holstein associations that developed the herd test, herd classification, sire recognition, and the Uniform Score Card. He chaired the Milk and Legislative Committee of the New York Holstein-Friesian Association throughout the 12-year legal battle over milk pricing. The successful conclusion of "The Guernsey Milk Case" was a major step forward for Holstein breeders. Prescott's career as a publisher spanned 63 years from 1913 until his death in 1976 at age 83.

Early in 1928, following the death of Frederick L. Houghton, the *Holstein-Friesian Register* was acquired, leaving the *World* as the only national publication of the Holstein breed from that date to the present. M. S. Prescott became the sole owner in 1932 after the death of Frank Price.

Through the years of the Great Depression the magazine was published as a biweekly, starting in 1932, and during the paper rationing in 1944 it went on a semimonthly basis, which has been continued ever since. During those years, until 1938, the editorial staff consisted entirely of Maurice and Bill Prescott.

The *Holstein World* has been blessed with many dedicated employees who had long careers with the magazine. William A. Prescott rejoined the staff as Associate Editor in April, 1919, upon his return from overseas service. He held this position for more than 40 years until his retirement in 1960. Maurice Prescott's career spanned 63 years, from 1913 until his death in 1976 at the age of 83.

Mabel C. Chase started her *World* career as a typist in 1921 before reaching her 17th birthday. She became M. S. Prescott's secretary and eventually appeared on the masthead as Editorial Assistant. Her duties included management of the office and supervision of the makeup of the publication. She died in 1966 after more than 45 years of service.

In 1938, W. Theodore (Ted) Prescott, son of M. S. Prescott, joined the staff as Associate Editor. His career as a Holstein journalist spanned 42 years, up to his retirement in 1980, and included the positions of Advertising Manager, Business Manager and Publisher. During his years as chief executive officer of the corporation, the company discontinued the printing business and contracted for

The World *staff in 1948. Seated are Bill Prescott and Maurice Prescott. Standing from the left: Fay Barney, Bob Hastings, Mabel Chase, Ted Prescott.*

the press and bindery work, paving the way for the modern four-color *Holstein World* of the 1980's.

In 1947 Fay C. Barney, an Iowa native and son of one of the pioneer Holstein breeders of the Midwest, W. B. Barney, became an Associate Editor following a lifetime of work with the breed in various capacities. He was a popular and widely traveled reporter of Holstein activities until his retirement in 1971 at the age of 80.

Robert H. Hastings, son of the founder E. M. Hastings, and son-in-law of M. S. Prescott, in 1948 moved up to the *World* masthead after a year as Associate Editor of the *New York Holstein-Friesian News,* published by the *World* company. Ten years later, in 1958, Theodore Prescott and Robert Hastings became part owners of the *World* with M. S. Prescott.

Throughout the 1960's and 1970's these second-generation members of the Hastings and Prescott publishing team worked together successfully, Hastings taking over responsibilities as Executive Editor, and Prescott devoting his energies to the changing production system and general business management. Hastings became Publisher of the *World* after Prescott's retirement in 1980, and chairman of the board of directors in 1982.

In October 1953, David L. Morrow became Associate Editor of the *New York Holstein-Friesian News* and soon moved up to a similar position with the *World.* An outstanding photographer and artistic designer of Holstein advertising copy, Morrow became Livestock Advertising Manager of the *World* in 1964, a responsibility he has handled with notable success for over two decades.

The name of another Associate Editor was added to the *World* masthead in 1956, Robert M. McKown, whose original duties also included editing the *Pennsylvania Holstein News.* Few journalists have ever been able to combine the editorial and advertising functions of a magazine as successfully as McKown and his title on the *World* masthead in 1984 bears this out: Associate Manager, Advertising and Editorial.

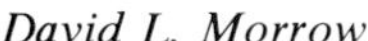

David L. Morrow

Robert M. McKown

James F. Hill joined the *World* staff as Associate Editor in 1964. Hill was on the *World* staff for 15 years and served as Editor of the *New York Holstein-Friesian News* for 10 years. Blessed with a photographic memory for Holstein cattle, he was particularly expert at organizing the annual All-American selections which involved pictures and show records of hundreds of winning animals annually (Chapter 9). Hill left the *World* staff in June, 1979, but continued to assist with the All-American program each year and also was involved in the preparation of this history.

Joel P. Hastings joined the *World* staff in June 1968 as Assistant Editor of the *New York Holstein-Friesian News*. In February of 1969 he became Editor of the *Pennsylvania Holstein News*, and in June of that year his name was added to the *World* masthead as Associate Editor. In January of 1978 he became Managing Editor, and subsequently became Editor and Publisher of the magazine and President of the corporation in 1980.

Under the leadership of the Prescott and Hastings families, the *World* has changed to meet industry needs. As state Holstein associations formed across the country, the *Holstein World* expanded with them, often providing the publication at reduced prices through the state groups. The Minnesota Holstein-Friesian Breeders' Association was the first to request a discounted subscription rate for all its members, to be paid by the Association and collected as part of their annual dues. "We made them a price of three for a dollar," recounted M. S. Prescott in memoirs published in 1979. From that beginning grew a system adopted by nearly every state association in the country.

This mutual effort has continued for over 30 years, although subscription at reduced prices is now optional. It was an important selling point for state association membership and also was of great value to the *World* in building and maintaining stable circulation.

The *World* has made many contributions to the development of the breed as Holsteins spread across the United States. One of the most important contributions involved integrity — integrity in livestock photography. Professional photographers had always retouched their pictures to make the subjects look as

Joel P. Hastings, Editor and Publisher of Holstein World.

attractive as possible. This practice began creating problems in the livestock industry around the time of World War I for the cattle market was hot and demand for pictures for promotional use was increasing. The first livestock photographers retouched every subject they photographed, filling in sagging backs and cutting off rough places. Retouching was accepted as a legitimate part of the business.

In the early days of the All-American contest, the *World* had a running battle with livestock photographers to get pictures that had not been altered, retouching being a common practice expected by exhibitors. The other breed publications took no stand on this issue but the *World* banned use of retouched pictures about 1920. The late Harry A. Strohmeyer, Jr., the foremost livestock photographer on this continent for a half century, about 1950 announced he would do no more retouching of animal photos in any breed. It cost him most of his beef cattle business, but that was not important to him as he was already overwhelmed with work in the dairy breeds. Following this precedent, today's leading dairy livestock photographers have long since renounced the practice of retouching animal pictures. This pioneering effort of Maurice Prescott of the *World* was a major factor in raising the level of ethics in livestock photography.

Since the days of Strohmeyer, photographers have been an essential part of the breed publications. More than half of the total magazine pages printed in breed journals today consist of carefully posed portraits of the best animals the breed has produced. Good pictures are the mainspring of livestock marketing, and marketing is what the breed press is all about.

The pioneer photographers were Robert Hildebrand, Harry Strohmeyer, John Carpenter, and Cook & Gormley's Livestock Photo Company. Following the dispersal of the great Dunloggin herd in 1943, Fred C. Wetmore, Dunloggin's herdsman, entered the picture-taking profession, serving all the dairy breeds until his retirement in 1970. Meanwhile other photographers served in smaller territories: Bill Serpa, Modesto, CA, on the West Coast; Harold Toles, Cobleskill, NY, in New York and Pennsylvania, (and he later sold prints from the extensive Strohmeyer & Carpenter negative file); Dick Green, Mason, MI, in Michigan and

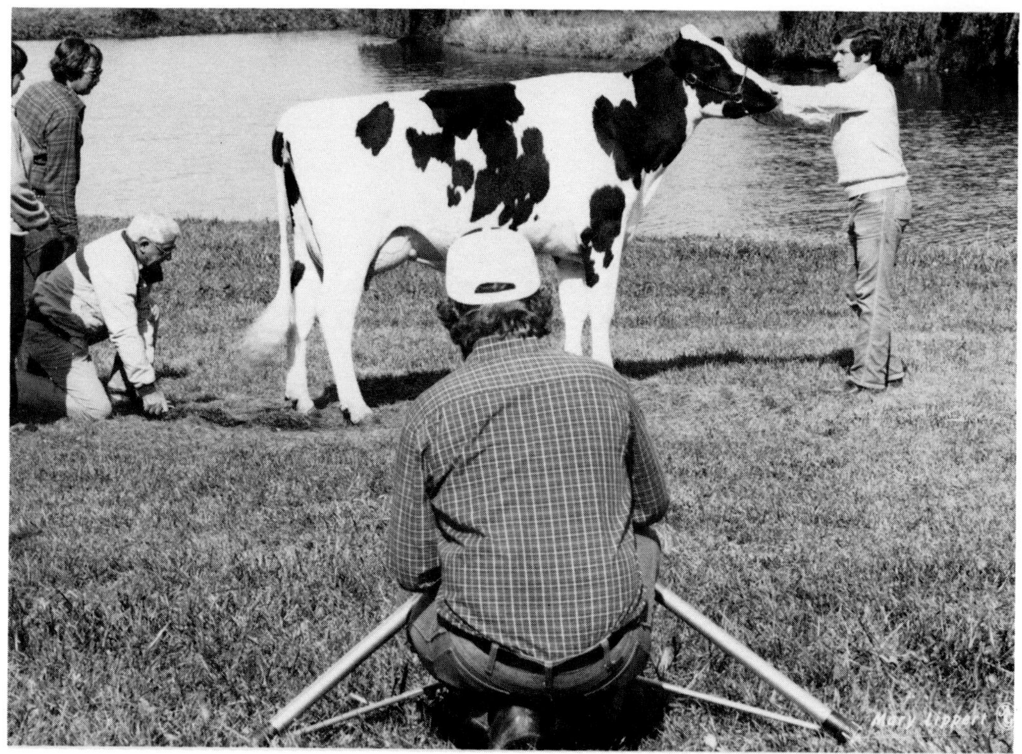

Photographers have been a key part of Holstein marketing for over half a century.

Indiana; J. H. "Jack" Remsberg, Middletown, MD, in the southeastern states; Pete's Auction & Photo Service, Wykoff, MN; and Mark Jenson, Walpole, NH, in New England.

In 1968, Danny Weaver, who had been employed in the AI industry, set up a new livestock photo service at Cary, IL — Agri-Graphics. He soon became so busy he began taking on associates, the first of whom was Jim Miller. In 1984 Agri-Graphics, Ltd. employed 11 photographers: Danny Weaver in Illinois; Mary Lippert, Eileen and Kathy DeBruin, and Tom Pearson in Wisconsin; Craig Johnson in Indiana; Fred Hall in Missouri; Maggie Murphy in New York; Cindy Taranto in Pennsylvania; Julie DeLavergne in Ohio; and Jim Fisher in Colorado. Also in the business in 1984 were veteran photographers Mark Jenson, Ken Vial (California), Jack Remsberg and Pete's Photo Service.

Many state Holstein associations publish magazines or tabloid newspapers (See Appendix). The first was the *Ohio Holstein News,* which celebrated its Golden Anniversary in 1984. The *Holstein World* has been actively involved in state Holstein publications since 1942 when it helped produce the *New England Holstein Bulletin,* now a slick bimonthly magazine. Two years later, the *World* began publication of the *New York Holstein-Friesian News,* and published the *Pennsylvania Holstein News* from 1955 through 1984.

Like the U.S. Holstein animal it chronicles, the *Holstein World* has become more efficient in the last two decades. In 1971, realizing that the printing operation was obsolete, Ted Prescott and Bob Hastings contracted out the work to Midstate Litho in Endicott, NY. With their modern web offset machinery and bindery equipment, the firm has handled the printing of the *World's* stable of Holstein publications for over 13 years, enabling the organization to publish well-designed modern magazines with sharp graphics and photography.

Although autonomous, the *Holstein World* and the Holstein Association have had a close working relationship for decades. One of the oldest joint projects has been the Honor List, the annual recognition of the breed's high producers. Calculated on a formula that considers both milk and fat production, the program singles out the top-producing Holsteins (in 1983, the top 2.8%) and is based solely on DHIR records. The Honor List began in the 1920's, after financial prizes for production were discontinued by the Association (Chapter 1). The list was published in booklet form by the Association until 1957 but even prior to that, the *World* published not only the names of the animals, but those of the owners, and the sires. The information was compiled manually until 1960 but now, computer listings are provided by the Association from which the special Honor List issues of the *World* are prepared. This ongoing cooperative venture has been popular with U.S. Holstein enthusiasts across the world.

A more recent joint venture between the *World* and the Association has been the Adventure Travel Series. Beginning in 1981 with a tour to Japan, the international tours have provided U.S. Holstein breeders the chance to also visit farms and shows in Australia and New Zealand, Europe (Holland, West Germany and France), and Great Britain. These trips promote U.S. Holstein genetics overseas and have built many a friendship among members of the worldwide community of Holstein breeders.

The *Holstein World* has grown as the breed has grown. It has matured into a magazine that is studied by Holstein breeders across the world. As U.S. Holsteins have gained popularity abroad, so has the *World.* It is the primary place where those in the Holstein industry look for information on what is happening in Holstein genetics, in the sale ring or show ring, or in the Holstein Association. *World* circulation in November, 1984, was 31,283 U.S. subscribers, 589 in

Cornell Ollie Catherine in the display window at the New York World's Fair of 1939.

Canada, and 1099 in 56 foreign countries on all the continents of the world. Japan and Italy, 284 and 134 respectively, led all other countries outside of North America.

Purebred Dairy Cattle Association

The New York World's Fair of 1939 had a display of five dairy cattle breeds — Ayrshire, Brown Swiss, Guernsey, Jersey, and Holstein — as part of the Borden's Dairy World of Tomorrow. The exhibit featured a milking merry-go-round called a Rotolactor. During the discussions that led to that cooperative showing, breed associations saw the need for some sort of a clearinghouse for ideas and action. An organizational meeting was held the following year with Glen Householder and H. W. Norton representing the Holstein-Friesian Association.

The basic plan of the organization was to provide representatives from each of the five major dairy breeds (Milking Shorthorns were subsequently added) and rotate the terms of office among them. Financing of PDCA was based in proportion to the number of transfers in the breed. All breed organization representatives would have an equal vote — there would be no individual members.

Through the years, PDCA has been an integral part of the development of dairy industry procedures and standards. While the initial efforts of the group involved public relations, with joint exhibits at national dairy shows in 1940 and 1941, PDCA soon went on to develop the Uniform Score Card for all breeds and helped to standardize the rules for production testing (H. W. Norton played a leading role as head of the testing committee). One of PDCA's major contributions (Chapter 3) came in the early days of artificial insemination during the 1940's, when it established the policies and set the level of ethics for the industry and developed a standard for AI record-keeping. Since then, the organization has been involved in setting industry standards for progeny testing and proving, genetic evaluations, blood typing, show ring and sale ring ethics, and matters of animal health and disease control.

PDCA has provided a mechanism for dairy breed organizations to focus on matters of mutual interest and concern and also has served as a contact point through which other dairy industry efforts can relate to the purebred segment. It has allowed the purebred breeds to speak with a coordinated voice.

The Holstein Association has been a major contributor to the limited budget of PDCA and has provided its share of leadership. Holstein related persons who have served as president have been Carl G. Wooster, J. Homer Remsberg, R. DeWitt Mallary, and W. R. Brooks. H. W. Norton served as secretary-treasurer from 1941 to 1944.

Whether as a forum to help solve the mutual problems of the purebred industry, or as a launching pad for cooperative dairy industry efforts, the Purebred Dairy Cattle Association has served the Holstein Association and the Holstein breed well.

PDCA members working on developing a Uniform Score Card inspect cattle during a field trip.

The Dairy Shrine, Fort Atkinson, WI.

The Dairy Shrine

In 1949, a small group of visionary leaders, headed by Joe P. Eves, developed the concept which led to the formation of America's shrine for the dairy industry. Since it was founded, Dairy Shrine has grown, developed and matured into an organization with more than 10,000 members located in all 50 states and many foreign countries.

The new permanent home for Dairy Shrine was officially dedicated at Fort Atkinson, WI, on September 29, 1981. The dedication address was given by Robert H. Rumler, Executive Chairman of the Holstein-Friesian Association of America. In his remarks he said:

"What you see today is the culmination of a dream shared jointly by industry leaders with vision and with direction who were personally as well as professionally dedicated, fully and completely, to the advancement of our great industry. Their dream evolved in 1949 at the time of the annual meeting of the American Dairy Science Association. It took form later that year with the organization of the Dairy Shrine Club, later to be officially designated as The Dairy Shrine.

"The Dairy Shrine was founded to:

1. Honor great dairy leaders of the past.
2. Recognize today's great dairy leaders.
3. Inspire dairy leaders of the future.
4. Record important events of dairy history.

"The dream of an organization dedicated to the past, the present and the future of the industry and to the establishment of a permanent home around which these national activities might orbit, was shared by names such as Fred Idtse, Floyd Johnston, H. W. Norton, Jr., Earl Weaver, Harold R. Searles, Leslie V. Wilson, Ed Estel, Joe P. Eves and Karl Musser.

"Many more contributed to the advancement of the Shrine and to the principles it has espoused over its history. Yet, in the milestone we reach today, two names must be mentioned.

"Without a doubt, it was the impending retirement of Dean H. H. Kildee in 1949 that triggered the idea that such an organization was needed. And so, it was at the first meeting of Dairy Shrine that Dean Kildee was recognized as its first Guest of Honor in what is now the dairy equivalent of a National Dairy Hall of Fame...

"The second name which must be mentioned is that of Joe Eves, who for many years was the driving force behind the growth and advancement of the Dairy Shrine. Joe's dedication to the Shrine made it what it is today..."

Members of the Dairy Shrine include dairy cattle breeders, scientists, educators, industrialists and students from all fields related to dairying. The lifetime membership fee is used exclusively to carry out the Shrine's primary purpose of "stimulating, inspiring, educating and recording." There are no further dues or fees.

Each year, a contemporary dairy leader is named Guest of Honor and a number of pioneers in the dairy industry are recognized posthumously. A dairy cattle breeder is also honored. Pictures and biographies of those honored are added to the historic collection at the Dairy Shrine Permanent Home and Museum at Fort Atkinson, WI.

Joe P. Eves was the first secretary-treasurer, holding those offices from 1949 through 1968. Succeeding him was Arthur W. Nesbitt, 1969-75, and still treasurer. Miles R. McCarry was secretary, 1976-80, followed by current secretary James M. Leuenberger.

There is a most distinguished list of Guests of Honor over the years. Those most closely identified with the Holstein industry include Fred Pabst, 1951; H. W. Norton, Jr., 1955; A. M. Ghormley, 1957; Harold J. Shaw, 1959; R. W. Jessup,1962; Maurice S. Prescott, 1963; Robert H. Rumler, 1976; and Ivan K. Strickler, 1983.

Presidents of Dairy Shrine, again limited to those with primarily Holstein connections, include, H. W. Norton, Jr., 1951; Harold J. Shaw, 1964; Robert H. Rumler, 1968; Donald V. Seipt, 1975; and Ivan K. Strickler, 1978.

Holstein winners of Dairy Shrine's Distinguished Dairy Cattle Breeder Award since its inception in 1973 are W. R. (Dick) Brooks, Louisville, CO, 1973; Donald

Babcock Test Exhibit at the Dairy Shrine.

S. Collins, Malone, NY, 1975; Wallace N. Lindskoog, Turlock, CA, 1977; Max L. & Max K. (Kip) Herzog, Petaluma, CA, 1979; Robert F. Thomson, Jr., Springfield, MO, 1980; Elroy Borgwardt, Valders, WI, 1982; and Edward A. Reed, Lyons, KS, 1983.

A specific objective of Dairy Shrine is to acquire pictures, histories, books and other records and memorabilia of those who have made notable contributions to the development of outstanding breeding herds and to the advancement of the dairy industry. The historic gallery at the permanent home includes pictures and records of noted dairy leaders as well as young people who have made conspicuous achievements in dairying. A highlight of the collection is the pictorial record of all national breed champions since the first National Dairy Show.

Dairy Shrine's work has expanded beyond the original purposes of recognition and commemoration. Cooperation with other dairy groups, particularly the dairy youth organizations, has been an important part of the program. Activities are carried out under the guidance of 21 directors representing all phases of the dairy industry. Directors are elected by the membership.

For 35 years, the Dairy Shrine has provided a unique opportunity for individuals and groups from the broad field of "dairying" to work together toward a common goal — the betterment of the dairy industry.

PART 3

Holstein Sires of Prominence

Introduction

Nowhere is the genetic development of U.S. Holsteins better chronicled than in the story of Holstein sires. The genetic gains on the distaff side of the breed have been just as dramatic, but the impact of a brood cow, due to her limited number of offspring, is difficult to assess across the breed. On the other hand, prominent sires have thousands of sons and daughters providing performance and type data which thoroughly and impartially documents their achievements. In this section, some of the Holstein sires that have made history are highlighted. Prominent brood cows are described throughout the text of the book.

Early sires were judged on phenotypic results, primarily their results in a single herd and in the show ring, since no reliable classification or production testing information was available. When artificial insemination first became widely used, genetic information was still scanty, and most of the prominent bulls of the time were known for their show ring (type) results. Only in the last two decades, as sire selection tools have become more reliable, have prominent sires had their reputation based upon genetics — on the production and conformation of their progeny. Therefore, this section is not a ranking of 20 prominent bulls, but rather the stories of some prominent sires during three eras of Holstein history. It illustrates the remarkable development of the breed during the last century.

How were these 20 bulls chosen? If a dozen veteran Holstein breeders were asked to give their opinion of the all-time top sires of the breed, there would likely be 12 different lists. Yet, a number of bulls would show up on everyone's list. The prominent sires covered in this section may very well be those sires.

Because of the dramatic influence of artificial insemination since the 1950's, most of the animals are of the post-World War II era. Earlier editions of this history cover the earlier sires that made such contributions to the breed: names like Hengerveld De Kol, King Segis, Sir Pietertje Ormsby Mercedes, and North Star Joe Homestead.

chapter 11

Famous Pre-AI Sires

In some respects, it was more difficult for a sire to "make a name for himself" in the days before AI. His sphere of influence was limited by the number and the quality of the animals in the herd and within a reasonable distance of the farm, as well as by the constraints of handling a mature bull. Still, the reputations of these herd sires have stood the test of time — including the onslaught of the fantastic genetic achievements of modern bulls with their thousands of milking daughters. It makes the achievements of these early sires all the more remarkable.

Only in the last 20 years have the reputations of prominent sires been based on genetics.

Johanna Rag Apple Pabst 346005
Born: 1-24-21

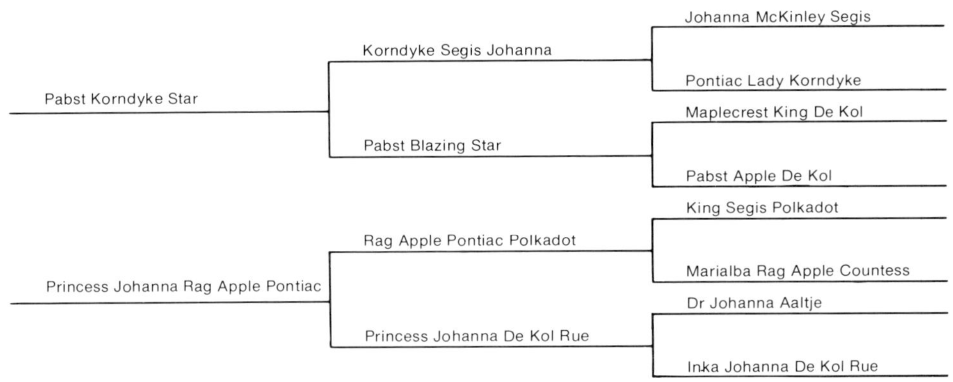

Johanna Rag Apple Pabst

The first Rag Apple, according to legend, was named after a rag apple bush or tree said to be growing on the farm of her breeder. Legend or not, history does record that a Holstein heifer, registered as Rag Apple No. 1286, was born on the farm of Don J. Wood of New York State, on March 6, 1885. This date, by coincidence, was just a few weeks before the committees met to lay the groundwork for the consolidation of the two warring factions of Holstein breeders, to form the Holstein-Friesian Association of America. The Rag Apple family, in its broadest sense, is also celebrating its centennial in 1985.

Some 36 years and eight generations after the birth of the original Rag Apple, Johanna Rag Apple Pabst (JRAP), was born on January 24, 1921. His contribution to Holstein history was as a show bull and a sire of top show type, along with high production and 4% test. Veteran breeders still refer to "Rag Apple" bloodlines when discussing direct descendants of Johanna Rag Apple Pabst.

JRAP got his Rag Apple genetics from the distaff side of the pedigree, his dam

Joseph Piek displays the Grand Champion banners won by Johanna Rag Apple Pabst at the 1924 and 1925 National Dairy Expositions.

being Princess Johanna Rag Apple Pontiac, with a junior 3-year-old record of 827.33 lbs. fat from 19,786 lbs. milk, 4.18%. Her sire goes back to the original Rag Apple on both sides, being a great-grandson of both Dutchland Sir Pontiac Rag Apple and of his full sister, Pontiac Clothilde De Kol 2d, the first 1000-lb. cow. Their dam, Pontiac Clothilde De Kol. was a maternal granddaughter of the first Rag Apple.

The sire of Johanna Rag Apple Pabst was Pabst Korndyke Star, who was a grandson of Pontiac Lady Korndyke (38.03 lbs. butter in 7 days) the highest short-time record daughter of Pontiac Korndyke. Pabst Korndyke Star is thus a great-grandson of Pontiac Korndyke. Another great-grandsire is King Segis, another of the breed founders.

Johanna Rag Apple Pabst was born on the farm of Philip Linker, a Wisconsin farmer-breeder. When the calf was about 8 months old, his dam having made a record of over 26 lbs. butter in seven days as a 2-year-old, he was sold to a neighboring farmer-breeder, Joseph Piek.

Johanna Rag Apple Pabst won his first show ribbon as a senior yearling at the Walworth County Fair, where A. C. Oosterhuis made him first and Junior Champion. Later that year, 1922, Mr. Piek showed his young herd sire at the Wisconsin State Fair. Well grown but shown "farmer-style," (without much fitting for the ring), he was soundly defeated, standing fifth in a class of seven — a rather inauspicious start for a history-making career.

His confidence unshaken, Mr. Piek brought JRAP out the next year as a 2-year-old, where he placed second at Wisconsin State Fair, was Senior and Grand Champion at Illinois State Fair, and topped the Waterloo Dairy Cattle Congress. The All-American Committee, consisting of W. S. Moscrip, R. E. Haeger and *World* Editor, Maurice Prescott, named Johanna Rag Apple Pabst All-American 2-Year-Old Bull of the year, over the National Dairy Show winner and 1922 All-American Yearling, Forsgate Mabel Ormsby Pete.

This began an illustrious show career starting in 1924 when he swept all four major shows he entered.

In the spring of 1926, Mr. Joe Piek was faced with a hard decision. Heading his small farmer-breeder herd was the top show bull in the nation. The first daughters of this bull were coming to milk, making creditable production records. Should he keep JRAP or should he sell him for a top price and carry on with some of his homebred sons and grandsons?

Col. O. G. Clark, the prominent sale manager who had staged the history-making John Erickson Dispersal in 1924, invited him to consign Johanna Rag Apple Pabst to the Clark Holstein Classic. It proved to be quite a sale. Bob Rasmussen, still in college, but with budding ambitions to build a truly great Elmwood herd, made his debut by purchasing Queen Bessie Pietertje Ormsby for $3300 and Max View Model Fayne for $3900. It was also the sale that sent Johanna Rag Apple Pabst to Canada—for $15,000.

T. B. Macaulay, a Montreal insurance magnate who was building Mount Victoria Farms, was trying to develop top Holstein show animals who would be high butterfat producers. He needed a herd sire, and after careful study, bought JRAP for $15,000, a new high for the period after the boom years of 1917-1920. It was a good investment: JRAP descendants, known as "Rag Apples," excelled in the show ring with exceptional udders and testing 4.0% or better.

Johanna Rag Apple Pabst was used on good cow families at Mount Victoria from the time of his purchase in early 1926 until his death in late 1933, which came as the result of an accident, when he broke his leg near the stifle. He was put away at 12½ years of age, just as his Canadian proving was getting well under way.

His highest record daughter was Montvic Rag Apple Colantha Abbekerk (EX) who set a new mark for the breed on 3X milking (1263 lbs. fat from 29,208 lbs. milk, 4.32%), was a Canadian National Grand Champion, and produced many

Montvic Rag Apple Colantha Abbekerk was the highest producing daughter of JRAP.

Montvic Rag Apple Bonheur was four times All-American.

superior progeny of both sexes including Montvic Rag Apple Sovereign and Montvic Rag Apple Marksman.

Montvic Rag Apple Abbekerk, with 886 lbs. fat from 18,121 lbs. milk, 4.89%, was the highest testing daughter of Johanna Rag Apple Pabst. She was the maternal granddam of Montvic Lochinvar.

His second highest daughter, Montvic Rag Apple Bonheur Abbekerk, was exported to the United States, and in the Manning Creamery Company herd in Iowa made 1047.4 lbs. fat with 4% test. She was full sister to the four-time All-American, Montvic Rag Apple Bonheur, who was dam of Montvic Pathfinder. Pathfinder, in turn, was sire of the 1941 All-American Get of Sire and two milking age All-American daughters.

Johanna Rag Apple Pabst, a bull of the era when the show ring set type and when production testing was in its infant stage, pleased many a judge, compiling a sparkling show ring record. But his prominence comes from more than his winnings. He was one of the few show bulls who transmitted both type and production to his progeny. His genetics helped build the Mount Victoria "Rag Apple" lineage that has had such an impact on both Canadian and U.S. Holsteins. JRAP was four times All-American, twice Reserve, and was picked as the Reserve All-Time All-American for the 1922-1942 period. He built a reputation that made "Rag Apple" genetics a part of the pedigree of most of the top U.S. Holsteins that would follow.

Governor Of Carnation

Governor of Carnation, the pride of Carnation Milk Farms during some of the most difficult years for the Holstein breed, is one of the all-time great sires in Holstein history. Famous for transmitting butterfat production during the pre-war period, when that was so important, he sired 580 registered offspring (307 males and 273 females). Governor was a bull of the selective AR testing era. Many of his daughters gained nationwide attention in the Carnation herd for their milk producing records.

Governor's pedigree is strong on both sides. His dam was Carnation Inka Walker Hazelwood, with 1149.4 lbs. fat, 4.7%, then known as the highest testing

Governor of Carnation 629472
Born: 5-21-30

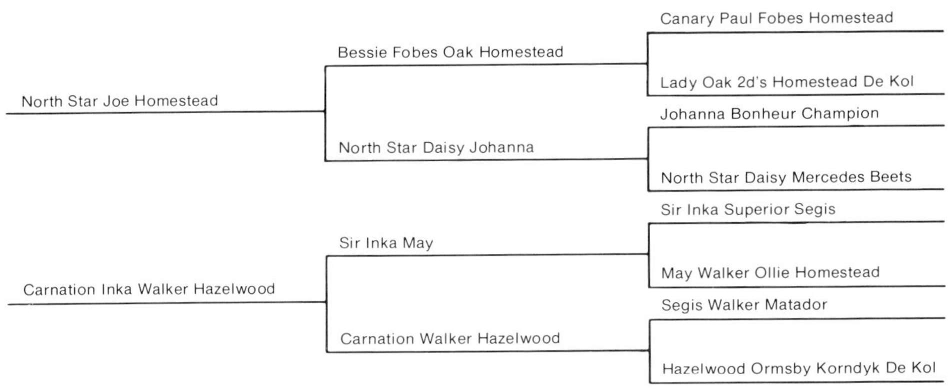

1000-lb. cow of the breed. She was a daughter of Sir Inka May from Carnation Walker Hazelwood, former world champion junior 2-year-old with 976 lbs. fat and later U.S. champion senior 4-year-old with 1198.9 lbs. fat, 4.1%. She in turn was a daughter of Segis Walker Matador from the old "Granny Cow" of the Hazelwood family, Hazelwood Ormsby Korndyke De Kol, generally considered one of the two foremost foundation cows of the Carnation herd.

Sire of Governor was North Star Joe Homestead, All-American Aged Bull 1924 and leading Honor List Sire 1930. Bred by W. S. Moscrip of Minnesota, he had been a herd sire in a small herd before being purchased by Carnation. He came to the main farm late in life and because of age and illness, his time in service was limited. But, he was successful in siring his greatest son, Governor, during his last years.

Bill Prescott, writing of Governor in the *Holstein World* in 1945, described how Carnation wished they had bought North Star Joe Homestead earlier. "I remember that day back in 1935 when I asked Rube Everly to let his hair down and tell me what he really thought of North Star Joe Homestead, and he replied: 'Bill, if we had him in the Carnation herd today in the prime of life and knowing what we now know about his ability, we would make him the greatest sire that ever lived.'

"That of course was impossible. Even Carnation cannot bring the dead back to life and virility. So they did the next best thing. They set out to make Joe's son, Governor of Carnation, the greatest sire that ever lived...

"They did this by mating him with as many as possible of their best cows and by testing and showing the resulting offspring."

As fast as Governor's daughters came to milk, they were run in the 3X class of AR testing, usually in their first lactation. Butterfat was the goal and they became known for it, setting many high fat marks. Governor had 38 daughters breaking the 1000-lb. mark.

The 4X testing program was popular in the 1930's and 1940's and Carnation used it extensively. Each year, a group of daughters was picked for the 4X test barn and Governor's daughters did well. His most notable daughter was Carnation Homestead Daisy Madcap, who in 1951, broke Carnation Ormsby Butter King's record for butterfat with 1413.6 lbs. on 4X milking.

Then in 1953, Carnation Homestead Daisy Madcap, already the champion of the 4X division, exceeded her own performance, this time on 3X milking to become the first 1500-lb. cow (1511 lbs.). Because of these results, credit must be given to this cow for providing the final proof of the wisdom of discarding the 4X division.

This record brought Carnation Homestead Daisy Madcap much publicity. Carnation Milk Farms made great plans for her; arrangements were made to bring her East as the star of their show herd. Nothing so elaborate had ever been

Carnation Homestead Daisy Madcap broke the world's record for butterfat in 1953 with 1511 lbs.

Carnation Homestead Inka Mutual set a milk record while on an extended show circuit.

planned before for publicizing a cow. A personal appearance tour across the United States, including TV and radio coverage, was in the offing.

But she had such little time to bask in the sun of her new glory, for on April 10, 1953, she died. She had freshened normally on the 9th; however, on the following day she showed signs of milk fever, but did not respond to treatment and died of a ruptured blood vessel before the day was over.

Another Governor daughter, Carnation Homestead Madcap, out of the 1313-lb. Madcap sister, set a junior 3-yr-old mark of 31,908 lbs. milk, 1216.4 lbs. fat, a record that was completed in 1941 and stood unchallenged for years.

Another noteworthy daughter was Carnation Homestead Inka Mutual, who was on test while making a show circuit of several thousand miles that brought her Grand Champion honors at such widely scattered state fairs as Wisconsin, Michigan, Indiana and Kentucky. Then, after showing at Waterloo, and dragging way down to Memphis for the National, she trekked back to Seattle to finish with 1248.2 lbs. fat and 32,470.4 lbs. milk. A couple of years later, under more favorable conditions, she made another record of 1333.1 lbs. fat from 34,681 lbs. milk, both years testing 3.8%.

The Governor heritage at Carnation was carried on by his four sons and four grandsons that were used in heavy service there. The sons included the three-time All-American, Carnation Governor Imperial, whose dam was a 1000-lb. daughter of Sir Inka May from the Hazelwood family. The other sons were Carnation Black Magic, out of the twice Reserve All-American Aged Cow, Ormsby Aaggie Tilda; Carnation Homestead Revelation, out of the 1174-lb. Grand Champion, Willowdale Skylark Beauty; and Carnation King Madcap.

Although Governor was a bull of pleasing conformation, there appears no record of his ever having been exhibited at any major show. The first record of any showing of his offspring was at the Pacific International in 1934 where his son, Carnation Governor Mutual, was first prize junior yearling bull and two other sons, Carnation Governor Imperial and Carnation Governor Butter King stood first and second in the bull calf class, with Imperial going on to become Junior Champion and later, being named All-American Bull Calf of 1934.

Progeny of Governor were very successful in the show ring through the pre-

World War II years. Gets of Governor were named All-American in 1939 and 1940.

Governor was a bull in the right place at the right time. In a famous herd known for its sires, he became known for transmitting high butterfat production. By the time of his death in 1945 he had compiled quite a list of accomplishments: he was the leading Honor List Sire in 1944; twice sire of an All-American Get; a "Century Sire" with 100 tested daughters; and his 25 Class 4X daughters averaged 1058.8 lbs. fat, a world record average. His sons and daughters and their offspring went on to many production records in the post World War II growth period, as many equated "Holstein" cows with "Carnation" cows. Governor has turned out to be one of the great pre-AI bulls of the breed.

Dunloggin Woodmaster

Throughout the early history of the U.S. Holstein breed there have been many examples of herd sires that, through their genetics, established the reputation of a herd as well as a breeder. Dunloggin Woodmaster is an outstanding example of such a sire. J. Homer Remsberg, former Holstein-Friesian Association President, put it like this: "It is my firm belief that Dunloggin, under the direction and management of that great cooperative team of Joseph Natwick and Paul Misner, contributed more to the development of the Holstein breed of cattle in America than any other herd in existence for an equal number of years."

Dunloggin Woodmaster transmitted production and type like few bulls of his time. On a dam-daughter proof, (56 daughters tested) he was +2000 milk. His PDT was +1.94 on 42 classified daughters.

But, unlike most of the other prominent sires of the breed, who were either carefully purchased (Johanna Rag Apple Pabst) or carefully bred (Round Oak Rag Apple Elevation), Woodmaster arrived as part of a package deal with his dam. Thirty years later, breeders would recall this when they heard the similar story of Pawnee Farm Arlinda Chief.

Dunloggin Woodmaster (GM) was born at Dunloggin, January 7, 1933. He came to Dunloggin in dam and the use of the Dunloggin prefix in naming him was in accordance with the bylaws of that day, even though he was actually bred by

Dunloggin Woodmaster 667915
Born: 1-7-33 Gold Medal

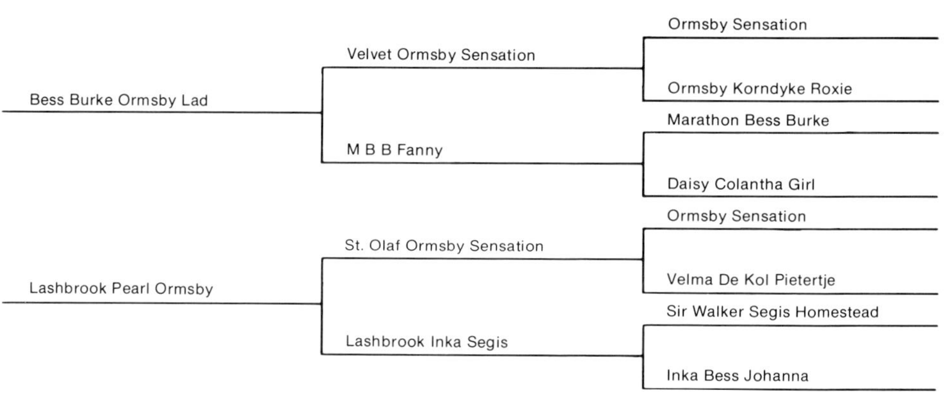

the Estate of Jerome H. Remick, Detroit, MI, a famed music publisher and president of Detroit Creamery Company.

Following Mr. Remick's death, the cattle were dispersed. Lashbrook Pearl Ormsby was a very highly regarded cow (she had a 971-lb. fat record), but she was 9 years old and thought not to be with calf. However, Paul Misner, Dunloggin's manager who was one of her greatest admirers, took a chance and bid her off at a low price. Luckily she proved to be with calf, and her son by Bess Burke Ormsby Lad, who was a grandson of Ormsby Sensation and Marathon Bess Burke, was named Dunloggin Woodmaster.

Dunloggin (Natwick had been in the lumber business and had cleared the logs on the property — he was "done loggin' ") built a remarkable herd around Woodmaster. There were four other sires before but his achievements overshadowed them all. Paul Misner, who would go on to be a noted cattle judge, used Woodmaster to build a herd that, when dispersed in 1943, would distribute fine seed stock animals throughout the breed.

Woodmaster had many notable daughters. Dunloggin French Mistress (EX-90) made the highest lifetime record of 267,714 3.8% 9662. In 1958, she was the only cow of any breed to have four records over 1000 fat at past 10 years of age. Her dam, Dunloggin Adowanna, made 888 fat. Dunloggin Happy Mistress (EX-92) became a national champion for milk on 3X milking with the highest record ever made at Dunloggin: 5-11 365d 3X 31,901 3.7% 1169. While French Mistress and Happy Mistress were Woodmaster's top producing daughters, two others stand out as his greatest daughters. Dunloggin Mistress Queen was the top female at the Dunloggin dispersal (Chapter 8) at $10,100 and sold to M. D. Buth & Sons, Comstock Park, MI. Dunloggin Proud Mistress sold for $10,000 to Curtiss Farms, Cary, IL. Proud Mistress was out of three-time All-American, Dunloggin Elenora, who was in turn from the All-American Aged Cow, Elenora Della Burke.

The Dunloggin Dispersal was one of the most publicized to this day. It was a tribute to Woodmaster, who, crippled with rheumatism, had been put to rest just days before. On September 15 and 16, 1943, the Dunloggin herd of 209 head, plus 21 baby calves, sold at auction for $383,700, and averaging $1835.88. There were 86 buyers from 21 states. Three animals sold in five figures and 18 animals crossed the $5000 mark.

Of this group of 18, 13 were sons and daughters and three were granddaughters of Dunloggin Woodmaster. He had 59 sons and daughters in the sale and they brought $152,675, for an average of $2587.71.

Woodmaster, product of a great cooperative team of owner, Joseph Natwick and manager, Paul Misner made the Dunloggin name synonymous with quality

Dunloggin Mistress Queen, shown here with her buyers, the Buth family, sold for $10,100 at the Dunloggin dispersal.

Dunloggin Mistress La Princess was purchased by Bob Rasmussen, Elmwood Farms. Dunloggin's herdsman, Fred Wetmore, who went on to Elmwood, is at the halter.

breeding. For decades, breeders following Dunloggin breeding were known for herds of uniform animals with good udders, sound conformation, and sharp dairy character. Woodmaster, by bringing Dunloggin to prominence, was not only a herd builder, but a breed builder as well.

Wisconsin Admiral Burke Lad

"Burke" is one of the sires who makes everyone's prominent sire list. He has had one of the most pronounced impacts on the U.S. Holstein breed of any sire, and more remarkably, did so before the era of artificial insemination.

Wisconsin Admiral Burke Lad was bred at Wisconsin State Reformatory, Green Bay, WI, and was born there on August 3, 1934. Archie Sandberg was herdsman and Glen M. Householder was farm advisor for the Wisconsin State Board of Control. Householder, the sparkplug in the development of purebred Holstein herds at the Wisconsin institutional farms, purchased their foundation animals mostly at Wisconsin auctions, and acquired, at the bargain prices of the Depression years, a tremendous asset for the state of Wisconsin. Householder also kept a tight finger on the breeding programs and matings in each of the institutional herds. He later went on to spearhead the Holstein-Friesian Association's Extension Program from 1938 to 1955. (Chapter 3).

In Archie Sandberg he had the advantage of a skilled cow man and keen student of blood lines working directly with the cattle. They used as many animals as they could afford which had the breeding of John Erickson, who was renowned for his breeding skill in using the genetics of Sir POM 37th.

Wisconsin Admiral Burke Lad was sired by a young bull bred at Green Bay, Wisconsin Admiral Burke. Intensely bred in the John Erickson lines, he combined with the Admiral Ormsby Fobes breeding of Emil Titel. There were 13 close crosses to Sir Pietertje Ormsby Mercedes in the pedigree of Wisconsin Admiral Burke, including four through Sir POM 37th and nine through Marathon Bess Burke.

In the Silver Glen Dispersal in 1932, Glen Householder purchased Burke's dam, Bell Farm Rosalind, who was a granddaughter of Colantha Johanna Lad. Rosalind had made two 30-lb. 7-day butter records, and at Silver Glen had made

Wisconsin Admiral Burke Lad 697789
Born: 8-3-84 Very Good-87 Gold Medal

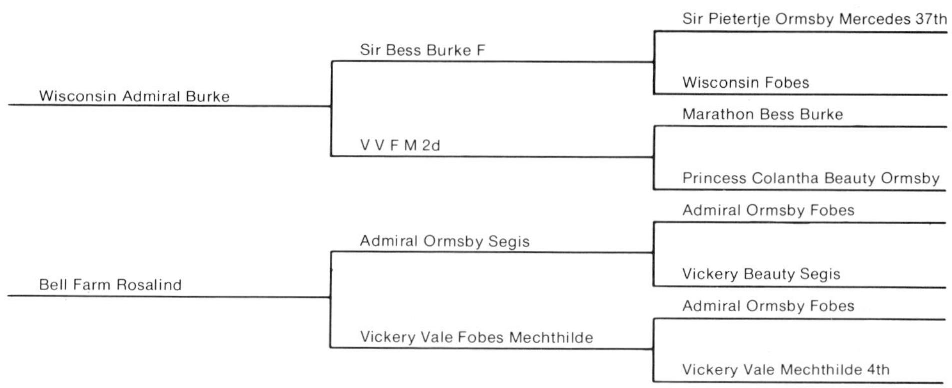

a year record of 19,758.9 lbs. milk.

The bull calf resulting from Burke and Rosalind was very appropriately named Wisconsin Admiral Burke Lad, from his sire, Wisconsin Admiral Burke and his dam's grandsire, Colantha Johanna Lad.

The young bull was sold to Utah State Industrial School, Ogden, UT, for the modest price of $150. By 1941 there were rumblings out of Utah that Wisconsin Admiral Burke Lad was a bull to watch. M. B. Nichols, then the Northwest fieldman for the Holstein-Friesian Association wrote in the *Holstein World* of March 29, 1941: "Mr. Walker states that not one daughter of this bull [Lad] has been inferior to her dam. One senior yearling daughter was Honorable Mention for All-American honors, this after having been Junior Champion at a number of western shows and the Pacific International. Several sons are outstanding show bulls and likewise siring the splendid dairy type and strong rumps that seem to be so prepotent in Burke."

Shortly after this, Wisconsin Admiral Burke Lad qualified as a Gold Medal Sire and with the highest production "index" of any sire of the breed at that time.

Pabst Farms was looking for a new herd sire. It happened that herd manager

Howard Clapp made several trips to Utah in 1941 before working out a satisfactory deal for Wisconsin Admiral Burke Lad. Burke left 30 sons in service in intermountain herds who made quite a contribution to the breed throughout that area. Many of them later found their way into other sections.

At Pabst Farms, Burke represented only a partial outcross. The Pabst herd carried strong Ormsby inheritance from the time of Creator (son of Sir Pietertje Ormsby Mercedes and Spring Brook Bess Burke 2nd). The impact of Burke on the Pabst herd and upon the breed was great although he was in active service there only a little over three years. In 1940 Pabst Farms had begun to use all their sires artificially and the influence of Burke accordingly extended more widely than would otherwise have been possible.

The Burkes made a distinguished showing in the All-American awards during the postwar years. Wisconsin Admiral Burke Lad won the Get of Sire award in 1947 and 1948 (also Reserve in 1950); his son Pabst Regal in 1949 and 1951; his son Pabst Roamer in 1950 and 1952, and Reserve All-American Get in 1949 and 1951; Wis Leader, a grandson of Burke had the All-American Get in 1956 and Reserve in 1955. Another son, Pabst Comet, had Reserve All-American Get in 1953, and grandson Weber Hazelwood Burke Raven had the Reserve Get in 1954.

The fame of the Burkes was great in the show ring, but they have also made an impressive record in production. Three sons of Burke were leading Honor List Sires: Pabst Roamer, Pabst Regal, and Pabst Comet.

Burke left a legacy that is unmatched in the Holstein breed. His genetics are spread through the registered population across the globe. Burke females were

Weber Burke Dusty Jo (EX-93-GMD) was Burke's highest classified daughter and first daughter to be a Gold Medal Dam.

Rocky Hill Mont Burke Dusty Jo (EX-95) was four times All-American and twice Reserve.

considered strong, wide, smooth, silky cows with good udders. They were typified by his most famous daughter, Weber Burke Dusty Jo (EX-93-GMD) and his best granddaughter, Rocky Hill Mont Burke Dusty Jo (EX-95), who in the early 1950's was four times All-American and twice Reserve All-American.

Most of the top sires of the last two decades, including Elevation, Astronaut, Burkgov, Bootmaker, Chief, and Kingpin trace directly to Wisconsin Admiral Burke Lad. A half century after his birth, his influence continues to be seen in the high producing animals of the 1980's.

Prominent Sires of the Early AI Years

As noted in Chapter 3, registered Holstein breeders were slow to use artificial insemination, fearing the loss of their market for young bulls. Then, as the technique gained acceptance, there were legitimate concerns raised over the possibility of betting wrong on a current "hot bull." Daughter/dam comparisons were still being used during the era of these sires; the evaluation tools were primitive for the demands being placed upon them. It was a transition period when the problems associated with using fresh semen also limited the influence of any one bull.

The animals described in this chapter are no genetic match for the great sires in the following chapter, yet each in his own way helped advance the Holstein breed in the post war decades. Many of the dominant AI organizations of the 1980's were built on the reputations of these fine animals. The seven bulls, known more for type than production, are thought by many to have sired the strength needed to support the milk production advances that were soon to come. These sires deserve their reputations — they earned them in a difficult, changing environment.

ABC Reflection Sovereign

ABC Reflection Sovereign, although used some in the Toronto District AI unit, is the last prominent sire covered in this book who was primarily a herd sire, rather than an AI bull. He was a Canadian bull but is included since his sons, including Marquis, and daughters were transmitters of type during a period of time when type was set in the show ring. He and his descendants have garnered over 100 All-American awards for their show ring success.

ABC Reflection Sovereign 198998C
Born: 11-29-46 Excellent-Extra

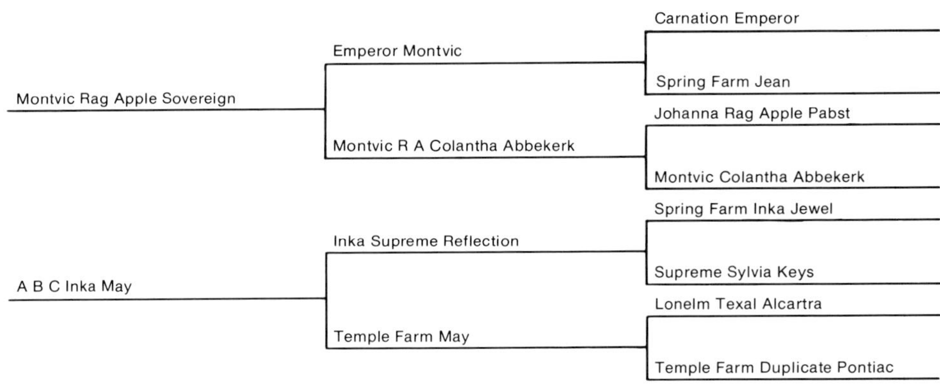

ABC Reflection Sovereign was an international sire. His pedigree included some of the best known Holstein names on either side of the border, while his offspring made history in many countries in addition to the United States and Canada. His dam, ABC Inka May, was All-Canadian 4-Year-Old in 1947, while his sire, Montvic Rag Apple Sovereign, had prominent sires on both sides of his pedigree (Governor of Carnation was great-grandsire and Johanna Rag Apple Pabst was grandsire on the bottom side).

Bred by Armstrong Brothers in Ontario, ABC was born in November, 1946. Two years later, Hector I. Astengo established Rosafe Farm, Brampton, Ontario, from the Armstrong's herd and their herd sire, ABC Reflection Sovereign.

To Rosafe owner Hector Astengo must be attributed the early recognition of ABC Reflection Sovereign's potential, and it was under the Rosafe banner that his early achievements were widely heralded.

ABC's mature lifetime almost spanned the years of Rosafe Farms' most active participation in Holsteins. The history-making Rosafe Dispersal was held on May 27th, 1958, less than a year after ABC's death.

Breeders of registered dairy cattle know there are many factors to be considered when evaluating the worth of a sire. When looking at ABC's contribution to the breed, a few of the "ground rules" under which he was used should be kept in mind. Early in life there were few, if any, unusual limitations on his service. Later, however, as his popularity became firmly established, new rules were promulgated, at least as far as his service in the U.S. was concerned. Under these rules, determined by his new owner, he was available only at a charge of $400 per service, and each breeder using him was limited to the registration of three offspring. Obviously, under these provisions, breeders using him at all did so only after considerable thought. The cows to which he was mated were carefully selected, and the resulting calves were prized, perhaps more than the majority of their stablemates.

The results of this careful selection of the dams of ABC offspring show up in the high level of these dams in the official U.S. comparisons for both production and type. Perhaps it is also evident in the fact that of his 32 sons with official U.S. proofs, 21 carried a Medal rating for production, type or both.

ABC Reflection Sovereign was a great "show" sire. He and his progeny had a remarkable record in All-American competition. The bull himself was Reserve All-American Aged Bull in 1951. One son, five daughters (one a repeater), plus five Get of Sire groups (made up of 13 different daughters), have been All-American winners, for a total of 12 All-American awards.

In the Reserve All-Americans, five sons, five daughters and two Get of Sire groups are listed. In addition, 11 sons and daughters and one Get of Sire group have won Honorable Mention and eight other sons and daughters have been

Rosafe Reflection Shamrock (EX-93-GMD) was ABC's first Gold Medal daughter.

Romandale Cora (EX-96) sold for $23,000 at the 1964 Romandale Sale.

chosen for nomination. This makes a total of 45 All-American, Reserve All-American, and Honorable Mention wins, and nominations by ABC sons and daughters.

In addition to the ABC record given above, paternal granddaughters and grandsons have also won All-American awards a total of 49 times and Reserve All-American 37 times. With other Honorable Mentions and nominations, the grand total in the ABC picture is well over 100 awards in the All-American program.

ABC Reflection Sovereign was the source of great show ring type, good production and the red gene for much of the breed. ABC sired the All-American Get of Sire in 1953-54-55-57-62 and Reserve Gets in 1960 and 1961. An ABC son, Rosafe Magician, sired the All-American Get in 1963, 1964 and 1965; and another ABC son, Rosafe Citation R, sired the 1966 and 1979 All-American Get of Sire groups. ABC and his genetics have been a dominant force in Holstein show rings for many years.

Wis Burke Ideal

Wis Burke Ideal rocketed to national fame through the outstanding production and show ring achievements of his offspring in the small herd of Elmer Dawdy of Salina, KS. At the time of his purchase by the Northern Ohio Breeder's Association (NOBA) in 1954, he rated as one of the top bulls of the breed based on the production and classification information on his offspring in Dawdy's fine herd. His 17 classified daughters averaged 85.8 with nine being Very Good and the other eight Good Plus. Early production information showed 13 daughters with an ME average of 16,694 lbs. of milk and 616 lbs. of fat. His offspring in the Dawdy herd continued to make further improvements for both type and production and were very successfully shown at some of the major shows. Interest in and use of the bull by breeders, not only in Ohio but throughout the U.S. and Canada, continued to grow as a result of his achievements.

Ideal was the result of a careful inbreeding program that sprinkled "Burke" throughout his pedigree. Wisconsin Admiral Burke Lad was bred to his own EX-92 daughter to produce Weber Burke Cyclone. Cyclone had a son, Wis Ideal,

Wis Burke Ideal 1013415
Born: 5-17-47 Very Good-88-Gold Medal

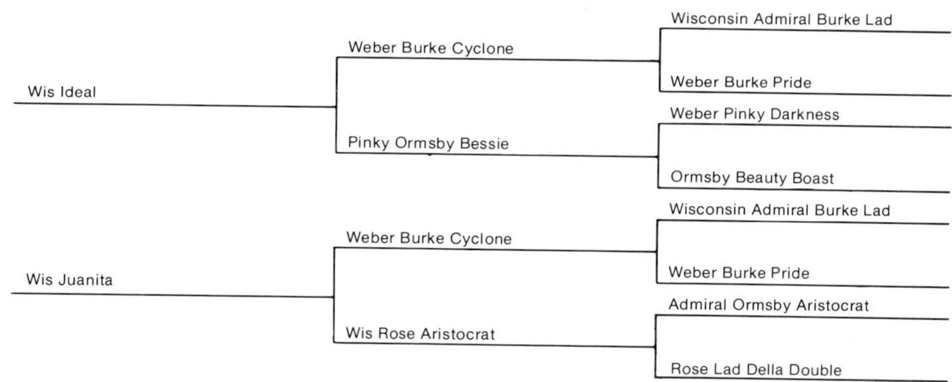

and a daughter, Wis Juanita, who were mated and produced Wis Burke Ideal.

Wis Burke Ideal was bred at Green Bay Reformatory, Green Bay, WI. He was picked for Mr. Dawdy by the Green Bay herdsman, Archie Sandberg, as one of the best sons of Wis Ideal, from one of the good-uddered families in the herd. Ideal followed Sir Bess Tidy (EX-GM), and Weber Hazelwood Burke Raven (EX-GM) and Raven Burke Hazelwooder (EX), in use at the Dawdy herd.

In a 1954 article in the *Ohio Holstein News,* Dawdy explained why he felt Ideal transmitted both type and production: "He's a double grandson of an inbred son of Wisconsin Admiral Burke Lad, combined with straight 37th breeding on the lower side of his pedigree."

Ideal had an illustrious active career at NOBA from 1954 to 1959. He was used for approximately 26,000 first services in Ohio. During a period when classification was just gaining acceptance and usage, he had 960 daughters classified with an average score of 83.0 (42 were scored Excellent). Farmer/breeders liked the milking qualities of Ideal's daughters, especially the persistence of production throughout lactation.

Get of Wis Burke Ideal at the 1954 Waterloo show.

Yet, in spite of Wis Burke Ideal's own record, the achievements of a world famous grandson helped anchor a place in history for him. Ideal was bred to his EX-91 daughter, Carine Mercedes Burke Ideal, to produce Tidy Burke Elevation (GM), whose son, Round Oak Rag Apple Elevation (EX-96-GM), became the number one sire of high performing daughters in the world.

Pabst Sir Roburke Rag Apple

Curtiss Breeding Service, Cary, IL, was one of the leading AI firms in the 1950's. One of the sires that brought Curtiss their reputation was Pabst Sir Roburke Rag Apple.

Roburke was born in late 1947 at Pabst Farms, Oconomowoc, WI, and was sired by Pabst Roamer (EX-92-GM). His dam was Pabst Burke Rag Apple De Kol (EX-92-GMD). Roburke was the result of a brother-sister mating — both parents were offspring of Wisconsin Admiral Burke Lad (Chapter 11).

At the age of six weeks, Roburke was sold from Pabst Farms to Mooseheart, Cornell Farm, and Neillewood Farm, all of Illinois. Later Merle Howard, R. B. Howard, and Donald L. Barkman each bought an interest in him. On January 2, 1954, he went to Curtiss, just a few days after receiving his Gold Medal Sire award.

Roburke had quite a career in the show ring and was twice Reserve All-American — as a senior yearling in 1949 and as a 2-year-old in 1950. In 1950 also he was a member of the All-American Produce of Dam. Three of his offspring have been All-American: Mooseheart Roburke Lucinda (EX), All-American 2-Year-Old Heifer, 1955, and Reserve 3-Year-Old, 1956; Mooseheart Pioneer (EX), All-American Senior Yearling Bull, 1958, and 2-Year-Old, 1959; and Mooseheart Roburke Ivy, All-American Heifer Calf, 1958. The 1956 All-American Produce of Dam, shown by I. N. and Ray Russell, Elgin, IL, were twin daughters of Roburke, and the 1956 get by him was Honorable Mention. The succession continued in later generations.

In 1960, the Holstein Association reported Roburke with 739 classified daughters averaging 80.9 points. In 1984, the report shows Roburke with 2120 classified daughters averaging 81.3 points and a PDT +1.33. He sired 37 Excellent

Rich Herd Roburke Matchless (EX-94) was Roburke's highest scoring daughter.

Pabst Sir Roburke Rag Apple 1024453
Born: 11-19-47 Excellent-92 Gold Medal 2/71

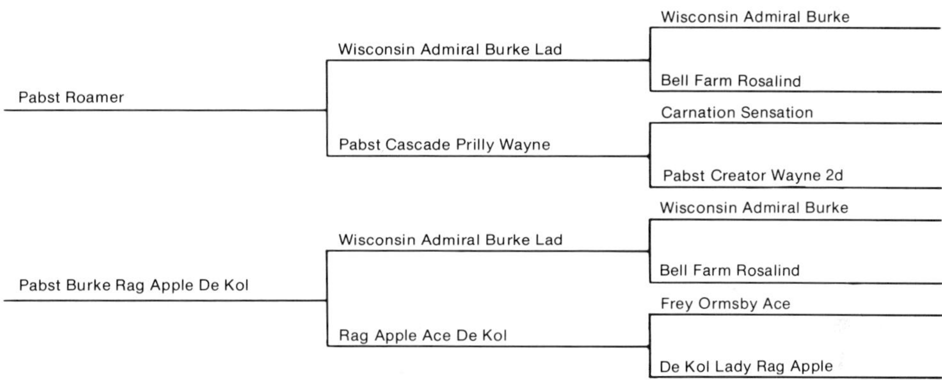

daughters, with Rich Herd Roburke Matchless being classified highest (EX-94).

In the way of production, the Association's most recent (1984) release shows Roburke with 230 daughters over 100,000M lifetime and 15 Gold Medal daughters. His highest milk production daughter, Verpana Farm Roburke Dot (VG-85), made 30,048 lbs. milk at 5-1.

Roburke was second Honor List Sire for 1958 and third in 1959 for production on his AR daughters, nearly all of whom were at Mooseheart.

Roburke had a full brother and two maternal brothers who were in leading studs. His full brother, Pabst Roamer Revelation (VG-88-GM), went to Consolidated Breeders, Anoka, MN; his maternal brothers were Pabst Raven Rag Apple 1236325 (GP) at Badger Breeders Co-op., Shawano, WI, and Pabst Sir Leader Rag Apple at the Quinte District Breeding Assn. in Canada.

At the time of his death in 1960, he had seven sons in the Curtiss stud. He also had 6000 registered daughters and 1929 registered sons.

Pabst Sir Roburke Rag Apple was a result of a brother-sister mating, with Wisconsin Admiral Burke Lad as his double grandsire. An illustrious sire in a noted AI stud, Roburke and his offspring carried on the "Burke" influence during the early years of AI.

Burkgov Inka De Kol 1038509
Born: 6-12-48 Excellent-92 Gold Medal

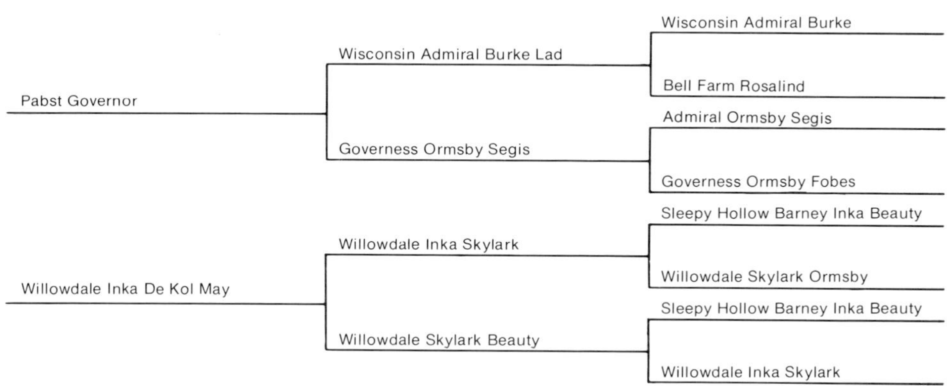

Burkgov Inka De Kol

A Utah dairyman once said of Burkgov Inka De Kol, "Eastman Kodak should erect a monument to him. He taught many a registered breeder to use a camera, as it was almost impossible to sketch his calves because of the spots."

This speckled white bull, who some top breeders would not use because of the black spots he passed to his progeny, soon made most people forget his looks and remember him for transmitting the type and production traits that breeders wanted. One of the first great AI bulls, he sired over 90,000 cows in the intermountain area of Utah and southern Idaho. During his peak year, 20,000 calves were born from his liquid semen.

John Stocking, sire analyst at Cache Valley Breeding Association, Burkgov's AI home, at the time, put it like this: "Holstein breeders soon recognized that the spots were dollar signs."

This episode in Holstein history began in the late 1920's when Holger Larsen of Dike, IA, returning from the Pacific International with a group of exhibitors from the Midwest, stopped at Montpelier, ID, to unload a young bull by the name of Forum Uniforum for Dr. Wallace Brown. Brown used the bull during the long

Depression of the 1930's, breeding some great animals that were either daughters or descendants of Forum Uniforum.

Wendell Fuhriman, herdsman at the Utah State University dairy, a man with a keen eye for a good cow, went into the Bear Lake country in 1941 looking for a good animal that could give Bill Walker at the Utah State Industrial School — with his daughters of Wisconsin Admiral Burke Lad — a little competition. Fuhriman found a diamond in the Brown herd, a cow called Willowdale Skylark Beauty, a granddaughter of Forum Uniforum sired by a Fred Eggers bull, Sleepy Hollow Barney Inka Beauty. The University paid $300 for Beauty.

Beauty had calved shortly before coming into the university herd. The following year, Wendell and Professor Caine, head of the dairy department, went back to see Dr. Brown and bought Beauty's yearling daughter, Willowdale Inka De Kol May, the dam of Burkgov Inka De Kol. Professor Caine, watching May develop into an outstanding cow, tried to breed her to her maternal brother, Carnation Homestead Revelation. This was in the days of fresh semen, and by the time the semen would reach Utah it would have no live sperm. The second choice was Pabst Governor, a son of Wisconsin Admiral Burke Lad.

Governor had come to the Utah State Industrial School as a result of the original commitment made when Wisconsin Admiral Burke Lad was purchased by Pabst Farms in Wisconsin (Chapter 11). Pabst Farms manager Howard Clapp had made several trips to Utah in 1941 in an attempt to buy Burke and, finally, an arrangement was made for Burke and four daughters to go to Pabst Farms in exchange for some cash, a bull for immediate use at the school, and later choice of a son of Burke from one of the good cows at Pabst's. The son selected was Pabst

Ideograph Burkgov Steps (EX-90-2E) was one of Burkgov's most famous daughters.

Ideograph Burkgov Inka (EX-94) was the highest classified daughter of Burkgov.

Governor. He was mated to Willowdale Inka De Kol May producing a spotted calf called Burkgov Inka De Kol.

One of the men at the university described Professor George B. Caine's reaction when he got his first look at Burkgov, "By hell, it looks like she got mated to a leopard!" Burkgov was leased to Cache Valley Breeding Association when he was but a youngster. He was used quite heavily in his younger days because of the respect of intermountain dairymen for his sire, dam, and maternal granddam. The dairymen of Utah and southern Idaho described the Burkgov daughters as members of "George B's new breed."

Like most great bulls, Burkgov was appreciated more after his death. His daughters and granddaughters developed into great producing cows — and many into outstanding brood cows. Following the genetics of such sires as Wisconsin Admiral Burke Lad and Forum Uniforum, he established a pattern of broad flat rumps and good quality udders.

In December 1958, when the NYABC (now part of Eastern) Holstein Sire Committee representatives checked Burkgov daughters in many Cache Valley herds, there were already 113 classified daughters averaging 82.4. The Utah-Wyoming AI daughter production summary covering 1954-1957 showed 325 Burkgov daughters averaging 13,365 3.5% 468, compared to the Holstein AI stud average of 438 fat. Twenty-five (25) HIR daughter-dam comparisons listed an increase of +2762M +86F. Burkgov himself was already Excellent and Gold Medal.

Burkgov AI daughters confirmed the impressiveness of their overall type. Udders were generally shapely; feet, legs, and pasterns appeared especially strong. Although upstandingness was not exceptional, body strength and rumps were good.

The recognition of Burkgov's greatness spread beyond the Cache Valley area. Bob Rasmussen of Elmwood Farms in Illinois purchased some top Burkgov daughters and used some Burkgov semen. His "Skokie" prefix appeared on one of Burkgov's best known sons, Skokie Glamour Boy (EX-91-GM), used in service at Select Sires.

Minnesota Valley Breeders Association had some of the first Burkgov sons used in AI studs. Peter Milinkovich, Minnesota Valley's sire analyst, later reported, "We feel Burkgov's influence has been greater for Minnesota dairymen through his sons than that made by the seven sons of Wisconsin Admiral Burke Lad previously used at Minnesota Valley Breeders. We also feel more Burkgov history will be written with each passing year."

In January, 1959, Eastern Artificial Insemination Cooperative expanded the parade of Burkgov breeding to the East. In addition to offering Holstein breeders Burkgov semen through its service program, many "selective matings" were arranged, and many Burkgov sons were added to Eastern's AI Proved Sire Development program.

Burkgov's reputation spread halfway around the world in the mid-1960's when 15 U.S. Holstein bull calves were selected for AI centers in Israel. The only sire with three sons in the group was Burkgov and after a young sire testing program was completed, and the unsatisfactory bulls were culled, his sons were three of the five selected. They were known for transmitting dairyness and good udders.

In the United States, as the AI industry grew, Burkgov sons and grandsons were used in most of the studs. In the late 1960's, AI stud sire selection personnel agreed that the influence of Burkgov Inka De Kol on the AI industry had been tremendous. It was doubtful that any other sire had his influence on dairy herds through artificial insemination during that period.

Ken Young, ABS Director of Dairy Cattle Breeding, said: "Burkgov would rank in the top of all bulls used in AI when considering both the high production level and high average classification. He has further contributed much to the industry through his many plus-proven sons in AI, as well as his top daughters being chosen for select matings."

The distinctive Burkgov spotted color pattern has become associated with many desirable type and production traits. This Utah-bred sire, a pioneer during the early days of AI, developed a reputation of transmitting good type traits, especially broad flat rumps and good quality udders. AI selection personnel liked him and his sons and at one point Burkgov had more sons in AI than any other sire. Four of the more popular were Skokie Glamour Boy (EX-91-GM), Sunnyside Standout (VG-85-GM), Tagalong Burkgov Dagan (GP-82), and

Burkgov Heilo Belle (VG-88).

Dick Chichester, Executive Vice-President and General Manager of Select Sires summarized the Burkgov influence this way: "Based on all facts, I truly believe that Burkgov was the greatest AI sire to come along. First, Burkgov sired top production. Second, he sired cattle that were 'made right' and lasted over the years."

Searsfarm Dean Ada Imperial

While Burkgov was leaving his genetic imprint in the West, another bull was becoming influential in the Northeast. Searsfarm Dean Ada Imperial, born in October, 1949 in the herd of James L. Sears, Baldwinsville, NY, was known for milk production traits and strong, fast milking, tight-uddered daughters. His daughters bettered the New York DHI average by 1282 pounds of milk and 43 lbs. of fat. While his ratings do not look particularly strong when compared to today's sires, he was highly regarded in his day.

Sears was sired by one of the Northeast's most popular AI sires of that era, Sir Bess Ormsby Fobes Dean (GM), the sire of many young bulls to enter AI service. Sears' dam, Searsfarm Resolute Countess Ada, was an Excellent-92 daughter of Pabst Sir Cascade Resolute. Her eight lactations averaged 20,995 3.3% 702 2X 305d ME and she had four daughters with 14 records averaging 1125 lbs. of milk above their herdmates.

Sears was sold as a calf to the Rondout Co-op. Bull Assn., Accord, NY. Rondout was a relatively small group of dairymen in the Hudson Valley maintaining a few sires to meet their herd breeding needs. In 1956, they expressed interest in placing Sears where he would have greater opportunity for extensive service.

Negotiations began and he entered the NYABC (New York Artificial Breeders' Co-op.) stud in Ithaca, NY, in May, 1956. At that time he had a January '56 AI daughter-dam comparison with daughters in seven Ulster County herds of:

19 dtrs. avg.	13,385M	3.6%	483F
19 dams avg.	12,980M	3.6%	464F
	+405M	-0-	+19F

Searsfarm Dean Ada Imperial 1085973
Born: 10-17-49 Gold Medal - 12/58

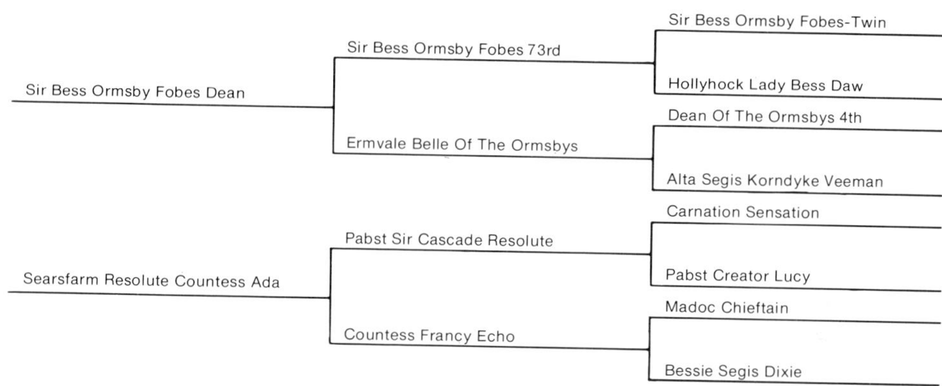

He achieved his Gold Medal rating in December, 1958. During his tenure at NYABC (May 1956 - August 1964), Sears was one of the most heavily used sires. Used primarily in the days of liquid semen, Sears was credited with about 265,000 services at NYABC. Looking at his eight-year record, his *monthly* average nearly equals the *annual* average of the typical AI sire during the period. He was extremely popular throughout the Northeast. Sears died on August, 1964 at nearly 15 years of age.

Sears' last published proof as an active AI sire was tabulated in January, 1964. It was based on 4500 daughters with 7485 records, and showed this superiority:

14,193M	3.6%	507F	AI dtrs. level
12,911M	3.6%	464F	NY Holstein DHI average
+1,282M	-0-	+43F	

Sears latest PDT, tabulated in July 1984 by the Holstein Association, shows +.86 on 3579 daughters. He also made an equal contribution to the Holstein industry through his sons. The latest sire-son report by Dr. Larry Specht of the

Pennsylvania State University dairy science extension department to include Sears was made in January, 1976 based on PD '74. It shows 73 total sons at +382M with 62 of the sons "plus." Thirty-three of them had 70% Repeatability or greater. They averaged +459M with 27 showing "plus."

His top daughters were:

Name & Owner	Age	Days	2X, Actual Production Milk	%	Fat
Colantha Searsfarm Ada (GP) Floyd Storch, Elmira, NY	5-3	365	33,540	3.6	1215
Chieftain D Elmore Sears (VG-87) Max Menendez & Sons, Walden, NY	5-4	365	31,013	3.8	1174
Ada Ormsby Imperial Ann (EX) Arlinda Farms, Turlock, CA	6-2	365	30,126	3.4	1035
Burke Imperial Prilly (VG) Colton Bros., Dalton, NY	6-7	335	28,490	3.9	1104

Searsfarm Dean Ada Imperial was extremely popular as a fine sire for production and type in NYABC country. Like his contemporaries, Ideal, Roburke, and Burkgov, his impact was regional in scope. Yet, during a period when the sire evaluation methods were primitive, Sears stands out as a good popular AI sire whose sons and daughters went on to advance the breed with the genetic progress of the 1970's.

Fleury Farm Dean Ada (EX-92-2E) was one of Sears' highest scoring daughters.

Ada Ormsby Imperial Ann (EX-92) was a daughter who combined type with high production (30,126m).

Osborndale Ivanhoe 1189870
Born: 4-26-52 Excellent-90 Gold Medal 9/73

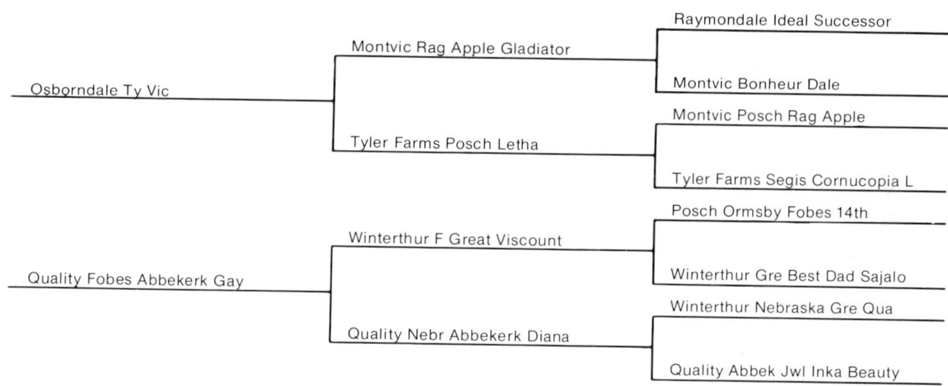

Osborndale Ivanhoe

Osborndale Ivanhoe came along at a time when many Holstein breeders felt the need to improve size, scale and stretch in their herds. An imposing figure himself — standing six feet, two inches at the withers and weighing around 2800 pounds — Ivanhoe was synonymous with siring these traits along with improved mammary system, dairy quality and good test.

Holsteins in the area where Southeastern Pennsylvania Artificial Breeding Cooperative (SPABC), later to become Atlantic Breeders Cooperative, was the dominant force in AI, were ideal mates for Ivanhoe in the mid-1950's: deep-bodied, strong boned, open-ribbed, but also sharp and dairy.

Ivanhoe's heritage traces back to some of the finest genetics in the history of the breed. His dam, Quality Fobes Abbekerk Gay (EX-GMD) was one of the great cows in the Osborndale herd at Derby, CT, and, with her family members, became the "female sensation" of the Osborndale Farm dispersal in 1957. Carrying 75 percent Winterthur blood, Gay was purchased by James M. Osborn of Whirlwind Hill Farm, Wallingford, CT. In this herd, Gay completed her lifetime

Professor Jim Osborn named Ivanhoe but passed up the chance to buy him.

totals of 192,519 lbs. of milk and 7428 lbs. of fat. Ivanhoe's sire was Osborndale Ty Vic (EX-GM), a bull with concentrated Montvic Rag Apple breeding.

Osborndale Ivanhoe was born April 26, 1952. Professor Osborn, who worked closely with Osborndale Farm owner Mrs. Waldo S. Kellogg, is credited with suggesting the name for him. As he later related in the *Holstein World,* Osborn passed up the chance to buy this future star sire: "...I had reserved the calf before birth and actually named him. The Gay cow was so big and so dairy, with two records over 1110 fat, and others of 999, 952, 932 and 807, that I eagerly sought to own her son. But when Ivanhoe was born he was such a skinny, scraggly calf that I relinquished my option to buy him. The moral: it's not a bull's type, but his genetic character which counts..."

Ivanhoe was used little in his early years in the herd and following Mrs. Kellogg's death in 1956, when the Osborndale herd was purchased by Charles Stroh, Hilltop Farm, Suffield, CT, ownership shares in Ivanhoe were divided at one-third each for Stroh, S. L. Bickford, and Aldo Panciera.

During those years, only Panciera and Bickford, along with partner Philip A. Coons, made full use of Ivanhoe, since the other owners had herd sires which they preferred over the youngster from Osborndale. Ivanhoe's heifers were rather rough in appearance, leggy, and awkward, but when they freshened their conformation improved and they showed potential to develop into great mature cows.

Like other bulls in this chapter, Ivanhoe helped make a name not only for himself, but for a stud. Sire Committee Chairman Earl Groff of SPABC was absolutely certain that this bull was meant to be in the Lancaster, PA, stud. Later, reflecting on the bull's success, Groff said, "He got us on the right road" to breeding better cattle.

Groff recalled traveling one Saturday in February, 1958 with Holstein Association classifier Jack Fairchild to evaluate SPABC-sired progeny on various farms. During the day, Fairchild mentioned seeing some impressive daughters of an Ivanhoe bull in Connecticut. On Monday morning, the sire committee was headed for New England.

At Bickford's Atlasta Farm, they looked at the bull and liked him, although others had described him as "too leggy and slab-sided." His milking daughters were very sharp and uniform, but were testing at a rather low butterfat percentage (which later improved). At the Tum-A-Lum Farm of Aldo Panciera, the committee was struck by the Ivanhoe daughters from the good-bodied but quite compact dams and their outstanding production and show ring performance. Size, scale, stretch, style, and tight udders were consistently evident on these early Ivanhoes. A 12-pair daughter-dam comparison showed increases of 2656 lbs. of milk and 102 lbs. of fat by the daughters, along with an average actual type score of 83.7 points. The committee concluded that the Ormsby breeding he carried was a logical addition to the SPABC stud at that time.

By March 4, 1958, a contract was drawn up to purchase the bull for $15,000, but with several conditions concerning health and response to semen collection. This was quite a high price at the time and the sire committee's action concerned the conservative members of the cooperative. (The total price was later lowered by $3000 through renegotiation with the sellers, due to poor physical condition of the bull which affected his semen-producing ability.) At Lancaster, an oversized stanchion was installed in his pen, his outside doorway was enlarged, and he towered above the fences that separated him from neighboring bulls. Fortunately, his temperament during handling was good.

Ivanhoe did not respond well to early collection attempts, but gradually improved and entered AI service in early April. An arthritic condition was soon diagnosed and heavy dosages of aspirin became part of his ration. Since this was before the widespread adoption of frozen semen, nearly all Ivanhoe semen was processed in its "fresh" (liquid) state for usage within a couple days around the SPABC membership area.

Entering the stud as a VG-88 and Silver Medal Type sire, he earned the Gold

Medal in March of 1959 with an official HIR production proof on 13 daughters that was very close to the original SPABC-calculated comparison of 12 pairs. When reclassified in August of 1962, he gained two points to go Excellent.

Ivanhoe's popularity grew rapidly and spread nationwide. Demand for his semen escalated. Unable to purchase semen, many breeders snapped up progeny at consignment and dispersal sales or purchased daughters out of Holstein herds in "Ivanhoeland," as the SPABC territory was nicknamed. Several large shipments were made to the West Coast in the early 1960's in response to breeder interest, especially to Edward Johnson of Diamond J Holsteins at Modesto, CA. Ivanhoe was the first AI sire to achieve national acclaim for both his semen and progeny, which helped other studs to realize the broad opportunities for marketing.

Despite efforts to provide special care, Ivanhoe's physical condition steadily deteriorated as his arthritis became more severe, and the semen supply dwindled. He went out of service in the spring of 1963 and died that November at the age of 11½ years. His remains were buried on the grounds and a marker was erected at the site. A very limited supply of frozen semen remained available for select matings, and it was not unusual to hear about transactions between breeders that involved several thousand dollars for one ampule.

As with the really great sires, the vast influence of Ivanhoe continues to live on through the performance by his thousands of progeny. He was the leading Honor List Sire each year from 1964 through 1971. Individual daughters and senior gets earned numerous All-American awards, and the top prices in countless sales were

Round Oak Ivanhoe Eve (EX-94-4E) was Ivanhoe's most famous daughter and the dam of Elevation.

Fultonway Ivanhoe Rae (EX-90-GMD) was Ivanhoe's highest fat record daughter (1615f) and the only cow with eight consecutive records over 1000f.

paid for his progeny. The "Ivanhoe influence" was chronicled for several years in special issues of the *Holstein-Friesian World.*

Ivanhoe genetics continue to be prominent in the pedigrees of top sires in AI service. Some of his most notable sons include Penstate Ivanhoe Star, Hilltop Apollo Ivanhoe, Provin Mtn Ivanhoe Jewel, Fleetridge Monitor, and Kilinsdale Ivanhoe Jack. Later popular sires with Ivanhoe close up in their pedigrees include Round Oak Rag Apple Elevation, Harrisburg Gay Ideal, Carlin-M Ivanhoe Bell, Glen-Valley Star, Arlinda Cinnamon, Eng-Amer Ivanhoe Jerry, Whittier-Farms Apollo Rocket, Wayne-Spring Fond Apollo, Poverty-Hollow Burkgov Demand, and Poverty-Hollow Milestone.

Ivanhoe was the great bull of the 1950's. Dairymen in the East were looking for the special traits he transmitted — taller, longer stretched Holsteins with tight udders. It is easy to see why Ivanhoe, despite a relatively short period of AI service, has been labeled by many as a "Father of the Breed." He transmitted both type and production. With a PDT of +2.53, he was the first Holstein sire to have over 5000 classified daughters (5687). He also had 622 daughters with records over 100,000M lifetime.

Osborndale Ivanhoe, the scraggly calf that grew to become a tower of stature, was at the right place and right time in Holstein history.

Romandale Reflection Marquis

Marquis, like his sire, ABC, was a "show ring" sire, known for his ability to transmit the strength and conformation favored by traditionalist breeders. Known by some as the "biggest moose that ever came out of Canada," Marquis was a magnificent physical specimen and one of the greatest of all show bulls. But he was a sire of type during a period when the emphasis of the Holstein industry had shifted to production, and thus is as controversial as any prominent sire of the breed.

Marquis was born at Romandale in 1956. His dam, Bonnie Lonelm Texal High (EX-5*), came to Romandale in June of 1953. Bonnie was one of five daughters of Lonelm Texal Highcroft (EX-ST) purchased for showing from J. E. Houck of Houckholme. They were remarkably uniform in their type and were potential get

Romandale Reflection Marquis 1459996
Born: 10-17-56 Excellent-95 Silver Medal Type

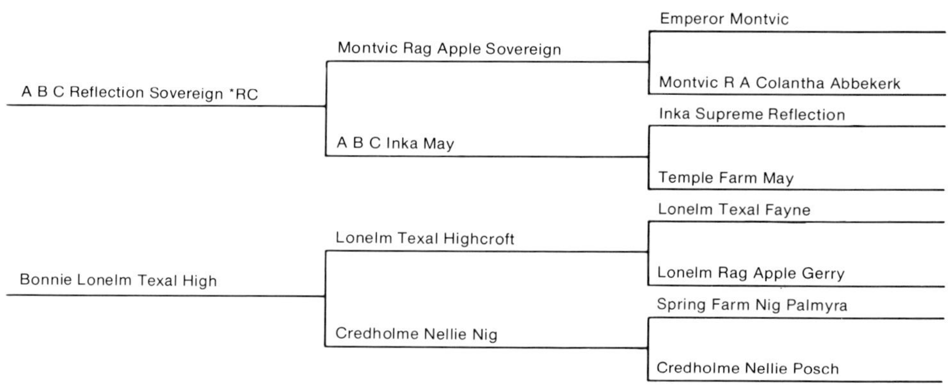

of sire. Bonnie was light in pedigree data since neither her dam nor maternal granddam had ever been tested or classified. Yet, she was a powerful angular cow that judges liked.

ABC Reflection Sovereign, Marquis' sire, was already known for siring show ring winners. It is no surprise that Marquis was a big gangly calf who took several years to attract attention in the ring. He won the 2-year-old bull class at the Royal Winter Fair. The following fall (1960) at the Dairy Cattle Congress in Waterloo, Judge Hilton Boynton made Marquis the Grand Champion.

It was surprising to many that Curtiss would decide to buy Marquis. He was a 7-year-old bull who had not been used extensively and had only 25 classified daughters. He had been three times All-Canadian and twice Reserve All-American and, while that was an advantage, no stud in North America normally would consider that sufficient reason for the purchase of a bull.

M. B. Nichols, after studying Marquis' milking daughters, came back from Canada in early 1964 with the suggestion that Curtiss should at very least consider that "big white show bull up at Romandale."

Nobody had ever talked too much about Marquis as a breeding bull, but

Brigeen Jenny Alert (EX-94-2E-GMD) was Marquis' highest scoring Gold Medal daughter.

M. B. Nichols suggested that the Curtiss AI firm consider "that big, white show bull..."

Nichols felt that he was a "natural" for use on the thousands of relatively small, but beautifully-uddered, "Burke" type cows in the United States at the time. Marquis could add the size and stature they needed and, in the process, make U.S. Holstein breeders more competitive in the show ring. Ted Krueger and M. B. Nichols were certain that Marquis, used on the right kind of mates, would do the right kind of job. History bears out their good judgment.

Curtiss bought Marquis at the 1964 Romandale Sale for $37,000. At the time, this was the second highest price ever paid for a Holstein at public sale in Canada.

Less than a week later, the "look at the size of him" legends concerning Marquis started to grow. He arrived at Curtiss by truck and Lawrence Kroner, the livestock manager, tried to lead him down an unloading chute. The bull was willing, but the chute just wasn't wide enough to accommodate his spring-of-rib. It never happened before and it never happened again, although hundreds of bulls of all breeds walked up and down that chute.

At Curtiss, Marquis was famous for size and power along with the eye-catching "balance" so important for the show ring. Already classified Excellent in Canada, he shortly picked up his 95-point Excellent rating in the U.S., going on to win his Silver Medal Type rating three years later, but with a substantial minus production proof in subsequent years. He and his progeny were very successful, and he became noted as the finest show ring sire of his day. He was All-Canadian Bull in 1959, 1962 and 1963, and Reserve All-American Aged Bull in 1962 and 1963. He sired 23 All-Canadians, 14 Reserve All-Canadians and 15 Honorable Mention progeny, as well as 29 All-Americans, 14 Reserve All-Americans and 37 Honorable Mention progeny. He sired three All-Canadian Gets and one All-American Get of Sire in 1970. Perhaps the greatest of all show bulls, he is as "controversial" a Holstein bull as has ever lived in the United States.

chapter 13

Great Sires of the "Genetic Era"

Sire selection techniques matured in the 1960's. Holstein Association Director of Research, Lynn P. Johnson, described the situation like this: "A lot of things happened at once. AI studs were pushing the program, extension specialists were behind it, DHIA was gathering the production data, and the Holstein Association was collecting the type trait information. All these facets of our dairy industry were acting in concert, recommending the same things, and dairymen began to use the tools. That all had to happen before genetic progress took place."

As the tools to measure genetics became more sophisticated, AI studs began to rank bulls by using predicted differences, choosing the very best sires of the breed. Competition was fierce and many smaller AI firms merged or fell by the wayside as breeders began to recognize that top quality sires had to transmit both type and production. Dairymen used the Sire Summaries — the results show it. As outlined by Robert Rumler in the Foreword, genetic development of the breed shot upward during this period. The new evaluation techniques brought a number of outstanding bulls to the top, sires that have accomplished things that the founders of the Holstein Association, in their wildest dreams, could not have believed. Here are the modern Holstein sires who have helped bring the breed to its premier position today.

Whirlhill Kingpin 1347940
Born: 2-13-59 Very Good-86 Gold Medal 1/83

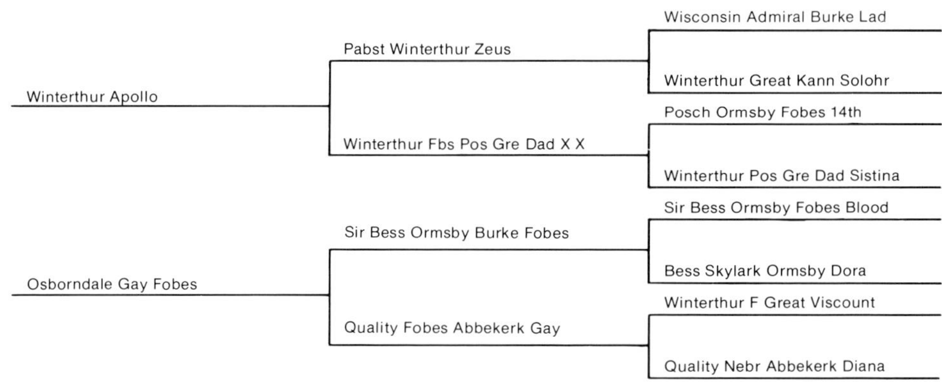

Whirlhill Kingpin

The catalog for the Whirlwind Hill Dispersal on May 28, 1959, revealed the high expectations bound up in the 3-month-old calf, Whirlhill Kingpin, who was listed just three spaces after his famous Excellent grandmother, Quality Fobes Abbekerk Gay. Today, it reads like a prophecy:

"Specially bred to be the next Whirlwind Hill herd sire, the result of six years of planning.

"Son of Apollo — from the daughter of Gay — an unusual specially created pedigree.

"The power of the Gay-Gwen cow family on the lower side of his pedigree blended with the overwhelming strength of the Christiana family behind Apollo on the sire's side, combines to make a truly dream pedigree for this sparkling youngster."

Kingpin, after a less than auspicious start, went on to become one of the top production sires of the breed. Living in the period when genetic measurement was just becoming sophisticated, he was still a generation ahead of the intensely selected sires of the 1960's. Yet, his record stands well against more contemporary

Kingpin was only 3 months old when he sold for $4000 at the Whirlwind Hill Dispersal.

sires with 36 daughters over 30,000 pounds milk (17th on list) and 51 daughters over 1000 pounds fat (16th on list). His results not only advanced the genetics of the breed, they helped build one of the modern AI firms — Sire Power.

Kingpin was bred by Jim Osborn of Whirlwind Hill Farm in Connecticut. He had purchased Quality Fobes Abbekerk Gay and her daughter, Osborndale Gay Fobes, at the Osborndale Dispersal (Chapter 8).

Gay Fobes, Kingpin's dam, possessed the qualities that Osborn was looking for, even though still just a 4-year-old. Her sire was the Gold Medal Sir Bess Ormsby Burke Fobes, purchased for $21,000 at the Osborndale Dispersal by a keen student of proven sires, Seward Johnson. Gay Fobes had gone Very Good at 2 years, had made 17,542 4.2% 712 at 3 years, and her first calf, Osborndale Admiral Lucifer, later well proven in Pennsylvania, had brought $2600 when only a 4-month-old calf. (She later made 22,310 4.0% 901).

The herd sire, Winterthur Apollo, was known for siring big, straight cows with superior udders and his daughters classified five points higher than breed average in both fore and rear udders.

Apollo's dam, the Excellent Gold Medal "Double X Cow," was a favorite of all visitors at Winterthur. Professor George Trimberger had picked Apollo's older full brother, Great, for the Cornell University herd where he earned a Gold Medal. Another full brother, Climax, proved out well for Winterthur.

Apollo died early from hardware. Charles Stroh, a well-known name in Holstein circles, has always bemoaned the early death of Apollo. He said many times, "If Apollo had lived and been given the wide service that Osborndale Ivanhoe achieved, I believe that Winterthur Apollo would have proven to be as great or greater than Ivanhoe." Since Charlie Stroh was co-owner of Ivanhoe at one time, his opinion carried weight.

Whirlhill Kingpin was purchased by the First Pennsylvania Artificial Breeding Cooperative of Lewisburg, PA, at the Whirlwind Dispersal at Wallingford, CT, on May 28, 1959.

The cooperative team had arrived at the sale planning to purchase another bull, but changed their minds once they saw the Gay family and outstanding cattle sired by Apollo. They paid $4000 for the calf, a price that was an attention-getter in 1959.

Kingpin entered service on February, 1960, at a year of age. Leighton Klingler, later a district manager for NEBA, reports that there was a great deal of interest in him, partially because of the price that had been paid for him. He was bred to everything, including a large number of cows with below average type. This was evident in his first type proofs, which showed that the dams of the classified Kingpin daughters were only 97.6% of Breed Age Average in classification score. Because of the influence of the early dams, the early Kingpin daughters were a mixed group and many dairymen had doubts about them. However, there was never any doubt about their production capabilities.

When Kingpin first entered service at First Pennsylvania, liquid semen was still being used. By the time he earned his proof, the First Pennsylvania, and its successor, NEBA, had converted to frozen semen so they were able to utilize 100 percent of the semen that he produced after he was proven, thereby enabling Kingpin semen to be available some years after his death.

Ironically, many of the better known breeders did not use Kingpin to any extent during this period. On the other hand, many other herds did, in many cases using Kingpin straight through the herd. As the resulting Kingpin daughters developed, they proved to be among the top, if not the top, cattle for production, type and sales appeal in these herds. As a result, the group of breeders who hadn't used Kingpin at first, began to, and later were the biggest group using the limited amount of Kingpin semen that was in supply.

Benbro Kingpin Polly (EX-96-3E-GMD) was Kingpin's highest scoring daughter.

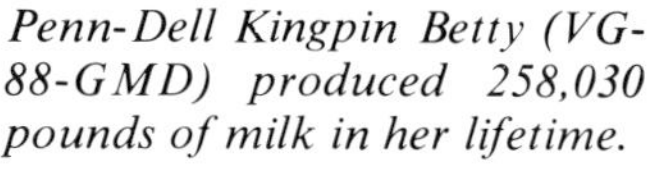

Penn-Dell Kingpin Betty (VG-88-GMD) produced 258,030 pounds of milk in her lifetime.

In 1964 the four co-ops, First Pennsylvania, NEPA, NJCBA, and Lehigh Valley merged to form NEBA, the Northeastern Breeders Association, with headquarters at the former NEPA site at Tunkhannock, PA. In September of 1964 Kingpin was moved to this facility and in the next four years bred nearly 120,000 first-service cows in the NEBA area, with his peak year being 1966, when he reached a total of 47,120 services. Prior to the formation of Sire Power, NEBA sold over 70,000 ampules of Kingpin semen to other organizations and breeders. Dr. Harold Schmidt bought a large supply which was sold on the West Coast.

Kingpin, the key sire in the formation of NEBA, proved to be the sire that opened the doors to NEBA's future relationship with other organizations. It is of interest to note that the first organization to obtain Kingpin semen in any quantity was the Maryland-West Virginia Bull Stud. Thus a relationship was established between these organizations that eventually led to the formation of Sire Power.

The Kingpin story illustrates two lessons: it is difficult to appraise the performance of a sire for type when he is used on below-average cows; and the contribution of the dam to the type of her daughter can not be overlooked. Kingpin's reputation, starting first for his ability to transmit production, grew as his first daughter was scored Excellent in 1970. Used with better animals, he achieved great results. With 7748 classified daughters (81.2 average), his current type proof is +1.45 PDT. He has had 235 Excellent daughters with seven scoring 94 or better. He had 322 daughters with over 100,000M lifetime and 35 Gold Medal daughters.

Jim Osborn, his breeder, once wrote, "To hold in trust the spark of genetic inheritance has inspired dedicated breeders wherever they farm." Kingpin is certainly a fitting tribute to that philosophy.

Ideal Fury Reflector 1381027
Born: 4-23-59 Excellent-94 Gold Medal 2/68

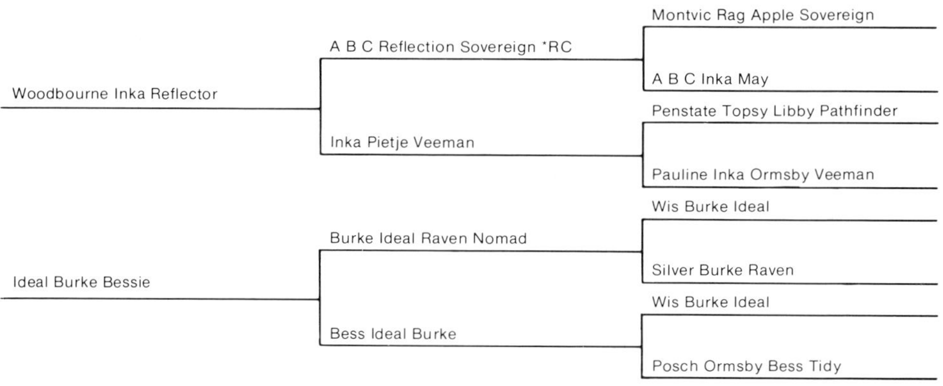

Ideal Fury Reflector

Ideal Fury Reflector was a specialist. He is the last prominent sire covered in the book to predominantly sire type, as did his sire and grandsire. His production proof was minus (PD -1488M) but he made up for it, in the eyes of his many fans, with his ability to transmit conformation. His success in the show ring is unmatched in the current era — he and his offspring have received 50 All-American nominations. For five straight years, 1971 to 1975, his gets won the All-American award and in 1984, his record was crowned with the selection of his 1972 Get of Sire as the All-Time All-American (Chapter 9).

Fury's pedigree is strong on type. Fury's current PDT is +2.93 on 3103 classified daughters. His sire, Woodbourne Inka Reflector, has a current PDT of +2.13 on 2181 classified daughters; and his paternal grandsire, ABC Reflection Sovereign, has a current PDT of +2.87 on 90 classified daughters. Seldom do three direct sires carry that level of PDT on nearly 5300 total classified daughters. His Wis Ideal (+1.16 PDT) and Wis Burke Ideal (+2.17 PDT) genetics on the bottom of his pedigree also contributed to his show ring prowess.

Jerry Conley, sales manager at NOBA, Inc., explained how Fury was sired:

"The Fury story really began with a Kansas farm boy doing chores for a neighbor. The farm boy was Irvin 'Spud' McDowell and the neighbor was Elmer Dawdy, one of the breed's greatest contributors to Holstein history.

"A typical Kansas cowpoke trade was consummated between Spud and Elmer. The deal was three baby calves from first-calf heifers for a specified amount of labor that Spud would do for Elmer. One of those calves was Ideal Burke Bessie, the dam of Fury. Dollar value for this heifer calf amounted to about $150. Bessie became Spud's 4-H calf and he showed her successfully until she was an aged cow.

"Spud was still working for Elmer when Bessie was a 2-year-old and together they loaded a full string of cattle in a railroad boxcar and headed for Waterloo. At the show were several milking daughters of Woodbourne Inka Reflector. Their great stature, length and full udders impressed Spud; thus he began to dream about the mating of this sire to his 4-H heifer.

"Of course with Reflector being a son of ABC, there was no way he could discuss this mating with 'Wis Burke Elmer.' However, upon return to McDowell Farm, Spud decided to carry out his dream and ordered semen of Reflector from a Pennsylvania source. The result was a bull calf born April 23, 1959."

Fury became the herd sire at the McDowell Farm, replacing Wis Blend, the fine AI sire from Wisconsin, in 1964.

Fury was a massive bull, especially in the fore rib and shoulder area. When he came to NOBA, he was scored EX-91, had 11 daughters with a production average of 18,869m 656f. They exceeded their dams by 1223m 39f. Of his 10 classified 2-year-old daughters that averaged 85.4, eight were Very Good and two were Good Plus. Combined they were 108.2% of Breed Age Average.

These daughters were extremely strong. They had wide chests and a near perfect blend of parts from shoulder to body carrying on back through to wide level rumps and firmly attached, correctly shaped udders. Fury was used heavily by Ohio and Missouri breeders, and his first champions were from these states.

It was the 1966 Holstein convention in Kansas that really got the ball rolling for Fury. Hundreds of people stopped by the McDowell Brothers Farm to see Fury's daughters. From that point on, Fury semen was in short supply. His

Gene Acres Felicia May Fury (EX-97-5E) was Fury's most famous daughter.

production proof at that time was on 15 daughters, all still in the McDowell herd:

15 daughters	19,402M	3.6%	695F
15 dams	16,375M	3.57%	584F
	+3,027M	+.03%	+115F

Nineteen classified daughters averaged 84.3, for a 104.7% BAA. In 1968 Fury daughters were in championship circles and beginning to hit the All-American pages, yet he only had 27 classified daughters that averaged 85 points in the July summary.

Fury died at 10 years of age, just five years after he came to NOBA.

Fury's most famous daughter was Gene-Acres Felicia May Fury (EX-97-5E). She was bred by Gene Acres Farm, WI and shown by Allen and Roy Hetts, WI. Felicia May was Reserve All-American 3-Year-Old in 1969, Reserve 4-Year-Old in 1970, and Reserve Aged Cow in 1971 and 1972. Then she won All-American honors in 1973 and 1974 as an aged cow. Felicia May was more than an All-American — she worked and lasted as she made 7-7 365d 3X 32,176 4.2% 1350 to be Fury's highest fat record out of a total of 4169 tested daughters. She also made the highest lifetime credits for butterfat of all the Furys: 227,830 4.3% 9369.

Fury's most famous produce were full sisters out of Sibu Ivanhoe Rebecca (VG-88). They were Rosemere Fury Becky (VG-88) and Rosemere Fury Iva (EX-95-2E). Together they were All-American Produce of Dam in 1973, 1974, and 1975. Iva was also All-American 4-Year-Old in 1972 and Reserve All-American Aged Cow in 1974. This produce was bred by William Shipley of Ohio and shown by E-L-V Apache Ranch, Michigan.

Fury's most famous granddaughter was Wapa Bootmaker Mandy (EX-96), All-American 2-Year-Old in 1975, who was sired by Bootmaker and out of a Fury dam.

Like Marquis, Ideal Fury Reflector made a strong impact as a sire of show ring type and he is prominent for siring type rather than production. Fury progeny received 50 All-American nominations, 11 All-American titles, 10 Reserve

All-American awards and five All-American Gets, a record that will be hard to break. Fury sired strength, stature, power and good udders, all valuable traits for the next generation of higher producing animals.

No-Na-Me Fond Matt

Loren Elsass spoke for breeders around the country when he said, "Cows that reach tremendous production levels, maintain their outstanding type, and then pass these qualities on to their offspring are a breeder's dream. No bull's daughters do this more consistently than those of No-Na-Me Fond Matt."

It's been over a decade since Fond Matt, a Kansas-bred son of Lakefield Fond Hope, died. In that period of time, he has come to be appreciated by Holstein breeders the way few bulls ever are. Known as a sire of superior type and high test, he has also earned the reputation as the sire of brood cows. His long-term contribution to the Holstein breed may well be through the sons of his outstanding daughters.

How did Fond Matt get such an unusual prefix? Marion McVay of Hutchinson, KS, who with her husband Hobart bred many fine Holsteins, explained: "It occurred to me that I had to always use McVay as a name and I had no real name anymore. So I chose that [No Name] and doubt if few people have ever thought why I chose it."

In 1944, the McVays sold their long-established registered herd, achieving the highest average and highest individual price in Kansas to that time. Retaining a few blemished old cows and heifer calves, they made annual trips to New York and Ontario to purchase animals. Among the herds the McVays visited was that of C. G. McKillican of Maxville, Ontario, who was breeding some fine cattle of Mt. Victoria lineage. It was here, in 1950, that the McVays were able to purchase an appealing white calf, Martha Maxview Rag Apple, who was to become the maternal granddam of Fond Matt. The calf had a strong Rag Apple pedigree, being a double granddaughter of Montvic Rag Apple Master with 12 crosses to Johanna Rag Apple Pabst.

In the McVay herd, Martha Maxview Rag Apple was bred to ABC Homer Sovereign, a good bull used on a limited basis mostly in Missouri, and at the time,

No-Na-Me Fond Matt 1392858
Born: 12-20-60 Excellent-91 Gold Medal 5/69

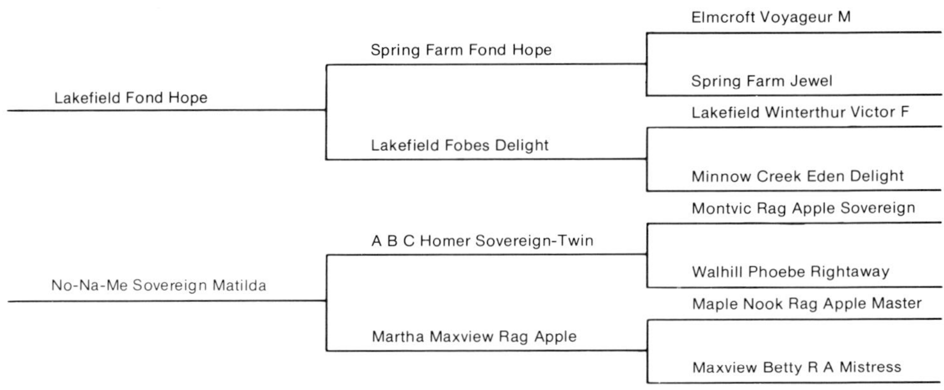

a McVay herd sire. The resulting calf was No-Na-Me Sovereign Matilda, who was described by Hobart McVay: "As a calf, Matilda was quite upstanding, and termed a bit leggy ... She developed into quite a cow, the tall sharper kind, with a good udder, and finally made the VG-88 classification."

Matilda was bred to Lakefield Fond Hope, then owned by the syndicate of which the McVays were members, and No-Na-Me Fond Matt was produced.

On a selection trip for the first COBA Invitational Sale in 1962, Dale Homer and Charles Baldwin were at the Marion and Hobart McVay Farm trying to pry one of their good Lakefield Fond Hope daughters away for the sale. While looking at the herd on a windy day in March, they noted the great old cow, No-Na-Me Sovereign Matilda (VG-88), by now showing some age. When they got back up to the barn area they saw this attractive young bull, No-Na-Me Fond Matt, who was Matilda's son by Lakefield Fond Hope (EX-93-GM), and made a deal for him.

Baldwin was quite enthusiastic about this beautiful calf. Fond Matt was in the first group of bulls for Select Sires' young sire program and was used considerably

Hobart and Marion McVay's "No-Na-Me" prefix is one of the most unique in the breed.

by the young sire sampling cooperators. Although his first daughters made a good showing, it was next to impossible to "sell him" to the distinguished breeders in the area. The Schneider Bros. of Kansas were among the first to use Matt because of the high regard in which they held Matilda and her daughters. One of their first daughters of Matt was Kanza Matt Tippy (3E-96), All-American Aged Cow, 1971, for Paclamar Farms.

After the second go-around with Fond Matt, his daughters began to really attract notice, and breeders started to use him heavily.

What really "made" Matt, and one of the great days of Charles Baldwin's life, was when Tippy became Grand Champion at the Central National Show to go along with her terrific production records.

From then on Matt received the recognition he so richly deserved. Charles Schneider outlined his success using Fond Matt at Kanza Holsteins: "We discovered what we considered to be an outstanding choice in a bull calf at McVays. He was No-Na-Me Fond Matt. We did not pursue the buying of this calf as we felt the $1200 asking price was too high. As soon as we learned Fond Matt

Kanza Matt Tippy (EX-96-3E) was the first Matt daughter to become an All-American.

went to COBA, we ordered nine ampules of semen at a cost of $1 each. Using it immediately resulted in six calves, four bulls and two heifers, all from Skyhawk dams. The two heifers turned out to be Kanza Matt Tippy and Kanza Matt Vi. Tippy eventually made All-American with over 36,000 lbs. milk and Vi (VG-89) had 32,000 milk 1200 fat. These were the first-born Fond Matts in Kansas and possibly in the nation."

Of the 26 Fond Matt daughters born at Kanza, five were Excellent and all eventually found their way to some of America's leading herds. Six Kanza-bred Fond Matt sons found their way to studs including Kanza Matt Tony who sired the Charity cow, the 1982 and 1984 Supreme Champion at Madison and All-Time All-American 4-Year-Old (Chapter 9). The Charles Schneiders became the breeders of Kanza Matt Tippy, Supreme Champion in 1971, and Kanza Matt Tony, sire of Charity, Supreme Champion in 1982 and 1984.

Those who believe that Fond Matt makes his greatest contribution as a sire of correct dairy type are amply supported by the record. He now has 7377 classified daughters that average 83.3 points, according to Volume II, 1984 Holstein Association Sire Summaries. The resulting Predicted Difference for Type (PDT) is +2.41 at 99% repeatability. This PDT level is the highest for any bull save two in Part III of the July 1984 Sire Summaries which includes the Top 200 TPI sires with 95% repeatabilities and listed as dead. The only two higher are Osborndale Ivanhoe and Oak-Hillranch Kit.

Fond Matt is the sire of 633 daughters that are Excellent. Of those Excellent daughters, there were 33 classified 94 points and above, up to 97 points.

But it was in the show ring that Fond Matt offspring brought the national spotlight to their sire. Kanza Matt Tippy's Grand Championships at the Central and Eastern Nationals in 1971 forced many breeders to take another look at this bull. Tippy exemplifies the sharp front ends, quality and dairyness that many Matts possess.

Gets of Fond Matt have an unmatched record in All-American competition. For nine out of the past 10 years, his gets have been either Reserve or Honorable Mention All-American. This includes a string of five consecutive Reserve awards from 1978 through 1982.

High fat test for his daughters is evident in the listings of sires with DHIR daughters over 1000 lbs. fat. Fond Matt ranks fourth on the list for 1000-lb. 305d records and 1200-lb. records, behind production greats Elevation, Chief, and Astronaut, all of whom have many more tested daughters. Fond Matt ranks fifth on the list of sires with 30,000-lb. DHIR records, again behind the three above plus Bootmaker.

Individual daughters have turned in extremely high production records. The list includes two daughters with over 40,000 lbs. milk and 1600 lbs. fat.

Matt ranks fourth in the breed for the number of Excellent daughters with 633 and fourth in the breed with 59 Gold Medal daughters. Matt's greatest achievement in the show ring would have to be that he appears as the grandsire of five of the six winners (great grandsire of the sixth) milking classes at the 1982 Central National which, of course, included the Grand and Reserve Champions. Premier sire of the event was the great Holstein Sire Development Service bull, Straight-Pine Elevation Pete (EX-90), proving again that behind many great bulls there is a great Matt.

Kanzabrook Matt Missy (EX-94-5E) was Matt's highest milk (42,040m) and lifetime (304,270M) record-maker.

Pawnee Farm Arlinda Chief 1427381
Born: 5-9-62 Excellent-94 Gold Medal 7/84

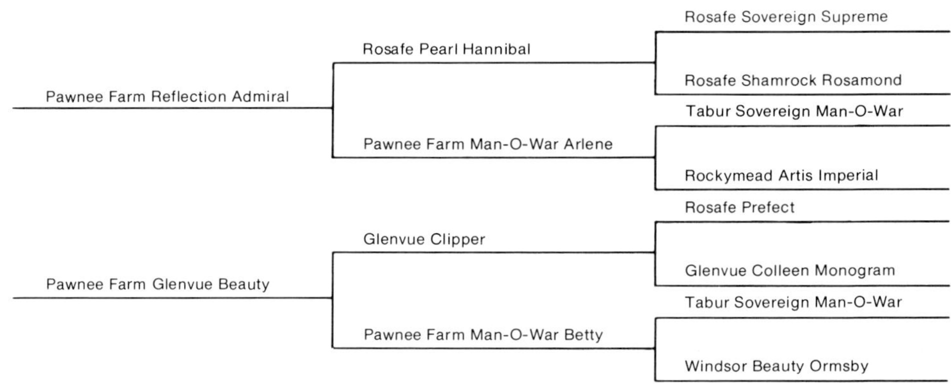

Pawnee Farm Arlinda Chief

While many of the breed's best sires are known for their milking daughters, probably none has a better record of top-ranked daughters and granddaughters than Pawnee Farm Arlinda Chief. His reputation for siring strength and will to milk is epitomized by Beecher Arlinda Ellen, who holds the world milk production record of 55,660 pounds. Yet, Chief came to W. N. "Wally" Lindskoog and Arlinda Holsteins only because they wanted his dam. It was one of the best "package deals" of the Holstein sale ring.

Lindskoog explained in a 1977 *Holstein World* feature, how he instructed Merlin Carlson, Arlinda Farm manager, to buy Pawnee Farm Glenvue Beauty at the Lester Fishler Dispersal Sale in 1962. Lindskoog said, "We have been asked many times if we bought Beauty because of the calf she was carrying or just because Beauty had what we were looking for. We bought the cow, and then thought about whether the bull calf born on May 9, 1962, was what Arlinda needed.

"We were looking for a careful balance between great will to work and great

Pawnee Farm Glenvue Beauty was carrying Chief when Wally Lindskoog bought her.

strength to work with — and we thought Arlinda Chief's pedigree gave him a good chance of possessing this ... the decision was to bring Arlinda Chief in as Arlinda's herd sire."

Chief came from a long line of Rag Apple breeding — primarily ABC Sovereign and Inka Supreme Reflection. There was some solid Ormsby influence through one of the breed's great "dairy bulls," Sir Pietertje Ormsby Mercedes 37th. According to Lindskoog, Tabur Sovereign Man-O-War, close up on both sides, gave Chief his ability to sire "those strong-muzzled, powerful, hard-working front ends that so typified the Arlene cow."

Chief was a hungry calf with a wide muzzle, broad chest, and enormous depth. Rus Shortridge, Holstein fieldman in California at the time, called it "a front end like a bulldozer." As Chief matured and developed, it soon became apparent that he passed this strength and ravenous appetite on to his daughters.

When the oldest Chief daughters started freshening, Joe Silva, then the master milker at Arlinda and a hard man for a cow to impress, told Lindskoog, "We have one of the all-time great milk bulls on our hands. I mean it, one of the all-time greats." He said this after milking only four Chief daughters for a period of just a few weeks, and the fifth one was just fresh.

The two men in the industry that, by their actions, first realized how great Chief was were Morris Ewing then of Curtiss, who updated Chief's track record monthly as the reports came in, and Doug Wilson of ABS who started right away to use Chief and his daughters for contract matings.

Chief was piling up a production track record at Arlinda that would be an almost impossible act to follow. Everyone wanted him for their stud. Curtiss, who had first refusal because of their earlier success in merchandising semen from unproven Arlinda bulls, snapped him up.

So, Chief came to Curtiss. Roughly a year after that, 1968, he came up with his first official production proof. His predicted differences were +1845 lbs. milk and +70 lbs. fat. Yet there were those who found them too good to be true. "Only 29% repeatability," a Curtiss distributor in Indiana pointed out. "And don't forget that those cows have been eating all that good California alfalfa. What's going to happen to Chief's proof when he gets daughters milking back this way?" At the

Chief's most famous daughter was the production record-holder, Beecher Arlinda Ellen.

time, few could believe that a bull's first 24 daughters could ME at 23,020 lbs. milk and 816 lbs. fat, a great feat still today.

At the same time, breeders weren't used to seeing 2-year-olds that milked the way most Chief 2-year-olds did. Dozens of Holstein breeders saw for themselves when they visited Arlinda Holsteins while in California for the 1969 Holstein Association Convention. Soon, as the AI daughters of Pawnee Farm Arlinda Chief started milking — really milking — all over the country, it was no longer necessary to go to Arlinda to see a Chief daughter. You could find them just about everywhere you could find cows. To see them was to want them. To milk them was to want more.

Chief, meanwhile, became a celebrity in his own right. He shared top billing at Curtiss with Paclamar Astronaut. Large numbers of people came to see both. Like Astronaut, he deserved his "stardom" with the results he achieved.

Chief's latest proof (July 1984) on 17,250 tested daughters in 5700 herds is +423 PD Milk +40 PD Fat +$96 PD Dollar +.14% PD Test +.02 PD Protein +.00% PD SNF. He is truly rare, being extremely high for test and still being used as a sire of sons, 22 years after birth. His latest type proof (July 1984) on 10,603 classified daughters is +1.53 PDT with an average age adjusted score of 81.8. He is off the charts for his most notable characteristic — *strength.* He sired a world of strength, wide rumps and wide udders.

Chief's son, S-W-D Valiant (EX-95-GM) is one of the highest TPI rated Holstein sires of all time.

Wapa Arlinda Conductor's nearly 500 proven sons rank first in the breed.

He sired some prominent sons. Wapa Arlinda Conductor (EX-90) and Glendell Arlinda Chief (EX-93) rank first and second in the breed with +100 proven sons as reported in the Penn State sire-son study of July '84. In fact, of the top eight sires Conductor ranks 1st; Glendell, 2nd; Betty Chief, another Chief son is 7th; and Chief himself ranks 8th.

Chief also accounts for 110 of the top 200 CTPI cows appearing on the 1984 Premier Performer list. S-W-D Valiant, a Chief son, had a decisive lead with 68 daughters, followed by Glendell with 21 daughters, Conductor with 10 daughters, Chief with eight daughters and Betty Chief with three daughters.

An article entitled "A Look at the Top 100 TPI Bulls" by Robin Drown appeared in the October 10, '84 *Holstein World.* Chief sired 12 of the top bulls while Glendell sired 11, Conductor sired eight, Betty Chief sired six, for a combined total of 37 of the top 100 bulls in the breed having Chief genetics.

The Chief chapter in Holstein history could be summed up this way — strong daughters that have plenty of will to milk. Chief is truly rare being extremely high for test (+.14%) and is still being used as a sire of sons. Chief sons — Valiant, Glendell, Conductor, Betty Chief, Memorial, and Mark — are powerful as sires of great cattle. Chief grandsons — Board Chairman, Enchantment, Misty, Tidal Wave, Jason, Fred, Bookie, Loue, and Spirit — are all top 20 TPI sires in the 1984 Holstein Association listing. For a bull calf that came to Arlinda unheralded in a pregnant cow, Pawnee Farm Arlinda Chief has cut a swath across the Holstein industry. At a time when genetic measurements were maturing, he made many a breeder a true believer in PDM. He left a legacy of strong milkers.

Penstate Ivanhoe Star

Penn State is known for great football teams, great dairy students, and a great bull named Penstate Ivanhoe Star. Recognized nationally and internationally as one of the most outstanding sons of Osborndale Ivanhoe, he is also one of the earliest random-sampled AI proven sires to make a significant and lasting contribution to the Holstein breed.

As this prefix indicates, Penstate Ivanhoe Star was bred by the Pennsylvania State University at State College. His granddam, Penstate Inka Veeman Anna

Penstate Ivanhoe Star 1441440
Born: 1-20-63 Very Good-89 Gold Medal 5/72

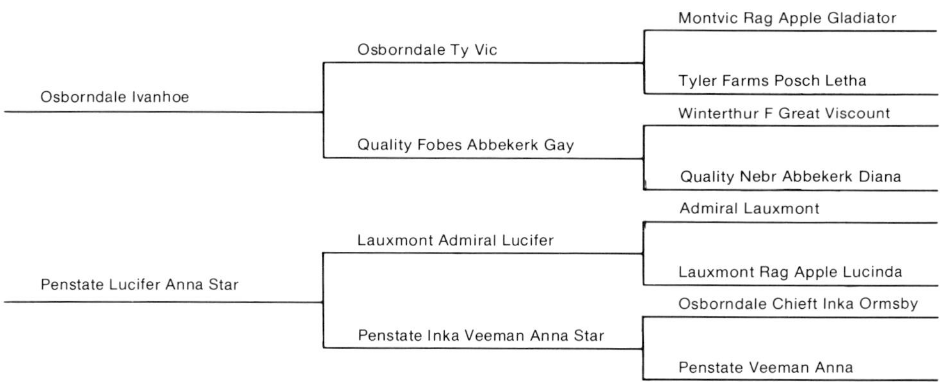

Star, was one of the favorites among students who milked cows while attending Penn State; she had a great disposition and was very easy to milk on the three-times-a-day herd schedule. During her lifetime, she produced 190,066 lbs. of milk with a 3.75% test and classified VG-87.

PSU dairy science professor Max Dawdy (later Executive Secretary-Treasurer of the American Guernsey Cattle Club) and dairy barns supervisor P. D. Jones bred her to SH10 Lauxmont Admiral Lucifer, an early stud patriarch at Southeastern Pennsylvania Artificial Breeding Cooperative at Lancaster, PA. This mating produced a tall, strong, very milky-looking heifer named Penstate Lucifer Anna Star.

Anna Star was admired for her dairy character, strength, and quality as she was growing. She was an immature 2-year-old, but showed potential to develop, and when she did, produced a string of 10 lactations that totaled 224,571 lbs. of milk and 10,173 lbs. of fat with an average test of 4.5%. Seven of her records surpassed 21,000m and five of them were over 1000f. Anna Star's highest record was her last, made at 15 years when she produced 25,612m and 1163f in 365 days to become the nation's senior fat leader. Classified VG-87, she is remembered as one

Penstate Ivanhoe Star was a tall, dairy calf with fine quality and style in the summer of 1963.

of Lucifer's greatest daughters and transmitted her qualities to several progeny in the Penn State herd.

When Anna Star was in her prime of life, PSU dairy professor Dr. Homer Cloninger was selecting the bulls to use in the Penn State herd breeding program. Osborndale Ivanhoe was at the top of his list and Anna Star, as an 11-year-old, was designated for a mating with this famous sire. The result was a tall, dairy bull calf with great quality and style, born on January 20, 1963 and named Penstate Ivanhoe Star.

As a 5-month-old calf, Ivanhoe Star was moved to the stud barns at Clarion, PA, and became known as WH103. He began his random sampling at about 14 months of age and proved to be very popular because of his pedigree.

Ivanhoe Star's first production summary showed only a small "plus" predicted difference for milk but those who knew the cow family in his pedigree were not surprised. They expected his daughters to develop rather slowly. R. Harry Roth, general manager of Atlantic Breeders Cooperative, at the time the manager and sire analyst at WPABC, recalled a story when the Ivanhoe Star daughters were just young cows. "A well-known, renowned Holstein breeder in the U.S. came to look at the Ivanhoe Stars as young cows. After observing them, he declared them to be just average cattle. But later he came back and looked at them again as mature cows. He admitted they had developed into great cattle as they got older."

Subsequent production summaries showed rapidly increasing milk PDs with good test (within three sire summaries, he was +1000 PDM, which was considered exceptional in 1971), while type data also improved steadily. A trademark of the first Ivanhoe Star daughters was the consistently tight udders tucked up under rangy frames.

High milk production, good test, size and scale, style, and good udders continued to be evident as they reached maturity and more daughters were evaluated. Ivanhoe Star was recommended as a mate for strong front-ended cows that needed more stature and improved udders. Several of his early daughters developed into cows with Excellent type scores, were very attractive in the sale ring, and moved on to other herds where they made important contributions to the herd improvement programs of other breeders. His popularity included

Great-View Mindi Star (EX-96-3E) was Star's highest classified daughter.

widespread use in commercial herds, many of them on the West Coast.

As an individual, Ivanhoe Star also developed gradually to attain the type classification score of VG-89 at 12 years of age. He earned the Gold Medal Sire award in May of 1972 and was among the top 10 Honor List Sires every year from 1975 through 1981. Over the years, his impressive lists of daughters with high Excellent type scores and records over 30,000 lbs. of milk and 1200 lbs. of fat have continued to grow. The longevity factor has also carried on, as more daughters are surpassing 200,000M in lifetime production.

On the international scene, he earned respect in Germany and Holland for the high protein level he transmitted. In Colombia, South America, Ivanhoe Stars were frequent Grand Champions at major Holstein shows, as well as placing high among cows on production test.

Ivanhoe Star died in April, 1977 at 14 years of age, completing one of the longest tenures of service by any sire in the history of AI. From mid-1964 until death, he occupied a pen in the Atlantic Breeders Cooperative barns at Lancaster. His availability continued through an inventory of frozen semen.

While the recent trend has been toward cattle that mature earlier, Penstate Ivanhoe Star rightfully earned recognition within the breed for other traits that are profitable to many breeders. In spite of the fact that his sons also sired slow-maturing cattle, his progeny also have developed with age.

Penstate Ivanhoe Star is sixth on the all-time list of sires with 30,000-lb. daughters with 95 (July 1984) and is also sixth on the 1200-lb. fat record list with 60 daughters. He sired 321 Excellent daughters and with 10,766 classified daughters, has a PDT of +.97. His influence has been carried on through several well-proven sons, including Carlin-M Ivanhoe Bell, Arlinda Cinnamon, Glen-Valley Star, and Eng-Amer Ivanhoe Jerry.

Penstate Ivanhoe Star will long be remembered for continuing the Ivanhoe tradition of tall, dairy, long lasting, good-uddered Holsteins, that became more popular as they matured.

Paclamar Bootmaker 1450228
Born: 9-22-63 Excellent-94 Gold Medal 7/84

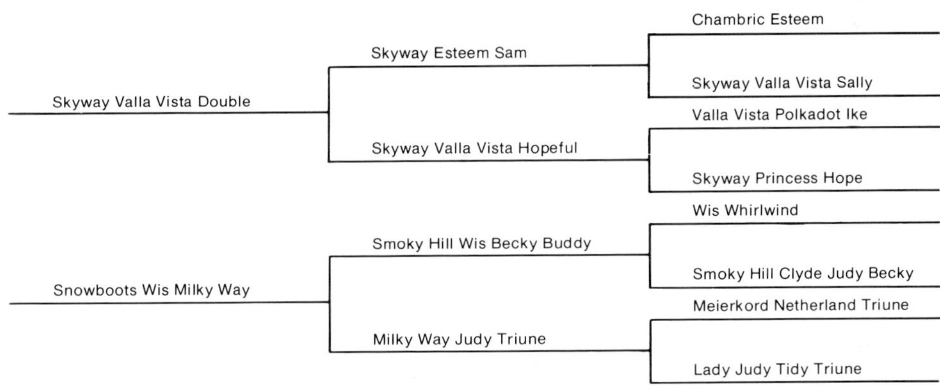

Paclamar Bootmaker

No bull in the history of the breed had ever sired over 100 Gold Medal daughters, until Paclamar Bootmaker came on the scene. Bootmaker, like Penstate Ivanhoe Star, is a sire that was on the leading edge of the genetic advances of the 1960's. The sire evaluation methods were just starting to become used intensively and his results show it. He was the son of a star of the show ring, Snowboots Wis Milky Way, and was well on his way to fame from just that heritage. But he has made his reputation as a sire of high-producing daughters, being just ahead of Ivanhoe Star with 145 daughters over 30,000 pounds milk. He ranks behind only Elevation and Astronaut with 17,501 daughters classified (PDT +1.17).

Bootmaker's dam, Snowboots, was a notable show ring winner of the 1960's. Classified at 97, she had a lifetime production of 201,187 3.8% 7729, yet she may be best remembered for her ability to gain affection and respect. Jack Sexton, who bought her as a baby calf, expressed it like this, " 'Boots' was one of the easiest cows to show; she was almost reluctant to leave the show ring. It was this style that was to later gain her national prominence. She seemingly dared a judge

Snowboots Wis Milky Way (EX-97-3E-GMD), a well-known winner in the show ring, was Bootmaker's dam.

not to put her on top. We happened to be at Waterloo when she won her class for the fourth time at a national show. When she pulled into the head of the line and set up, it was with that same style we had seen before, standing in the pasture a quarter of a mile from the barn."

Dick Brooks, who has become one of the most successful Holstein breeders in the "Genetic Era," purchased Snowboots as a young cow. He explained in 1976, that she was an intensely bred Ormsby and that he was looking for a strong Ormsby-pedigreed sire: "After I had worked with Boots for about a year and a half, trying to decide what the cow needed most, I decided to find a bull that I felt had the qualifications, and at the same time had the strongest Ormsby pedigree to unite with hers. I'll have to admit, I was trying harder at that time to stay in one bloodline than I am now. I still find that concept interesting but not the most important principle in breeding great cattle.

"After much thought and looking I decided to breed her to Skyway Valla Vista Double, a bull whose mother had the same width of rump as Boots, a wide high rear udder, and a lot of strength in the front end, both characteristics which she needed. He was extremely strong in Ormsby bloodlines as well."

The calf born in 1963 was named Bootmaker because Ted Prescott of the *World* asked during a visit whether Snowboots would be making a lot of little "Boots." Brooks decided to sample the young bull and also used him heavily in the Paclamar herd. When Paclamar dispersed in 1967, it was still early in the young bull's life and his record was not yet established. Because the bull was not completely proven and his fresh daughters were not necessarily out of the right kind of cows, the bull was not appreciated to his full capacity by the AI industry. Bootmaker was purchased by a syndicate headed by Ray Umbaugh, a new breeder starting in the business. Sam Fletcher of Ft. Wayne, IN, who had bought and proved Paclamar Double Triune, was another member. Several months later, the Bootmaker Syndicate leased Bootmaker to American Breeders Service where he lived until his death in January of 1976.

As the Bootmaker daughters calved all around the country and ABS began to compile their records, it was quickly apparent that his daughters could milk and keep at it. In January, 1969, ABS computed a 32-daughter proof that had

Wapa Bootmaker Mandy (EX-96-3E-GMD) was All-American 2-Year-Old in 1975.

Bootmaker at +732M and +21F. The rest of his production story is history. With a PDM of +266M and PDT of +1.17, he is exceptional for a sire of the late 1960's.

Bootmaker's daughter, Fleetridge Bootmaker Dixie (EX-90) was his highest milk producer with 6-7 365d 43,430 3.1% 1346.

Today, Leadfield Columbus-ET (VG-88) is Bootmaker's leading son for TPI. He is sixth among all AI sires ranked by TPI and is with ABS. Paclamar Triune Complete (VG-89-GM 5/72) also made a significant contribution as a Bootmaker son. He was available from Atlantic Breeders Co-op. in Lancaster, PA. A son of Complete, Carnation Glenview Joe-ET (VG-85) ranks 37th on the current TPI listing and is available from Carnation/Genetics in California. It is clear that the influence of Paclamar Bootmaker on the Holstein breed will continue.

Robert H. Rumler, writing in the Foreword, noted the dramatic genetic advances in the U.S. Holstein breed that began in the late 1960's. Paclamar Bootmaker, with 29,230 daughters averaging 17,688m, was responsible in part for turning the charts upward. Bootmaker daughters were noted for silky hides, good fore udders, angularity, a clean bone, and tremendous desire to milk. At ABS Bootmaker became the first Genetic Special bull, a category of special merit for unproven sires. He proved to be just that — a special merit Holstein sire.

Fleetridge Bootmaker Dixie (EX-90-2E) was Bootmaker's highest milk record daughter with 43,430 lbs.

Paclamar Astronaut 1458744
Born: 1-19-64 Excellent-90 Gold Medal 5/69

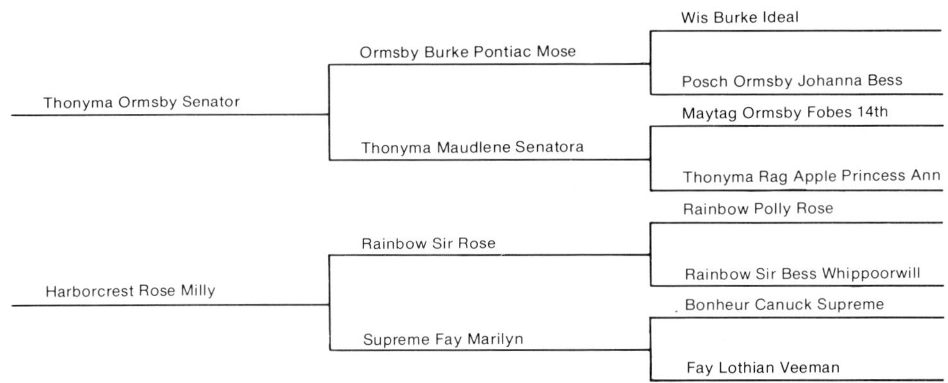

Paclamar Astronaut

"The only son of an Excellent-96 All-American 1000-lb. fat cow ever offered at public sale," stated the 1964 Maryland National Convention Sale catalog. The son offered was, of course, Paclamar Astronaut, born January 19, 1964, and sold for $9000 at that sale on June 11, 1964.

Dick Brooks of Paclamar Holsteins in Colorado, and a past president of the National Holstein Association, gave the details in a feature story for the April 10, 1976 *Holstein World:*

"I'm sure the Astronaut story is a mystery to everyone and I feel very proud at this writing to have our prefix on a bull with as much going for him as he does. Very few bulls have had as much use as has Paclamar Astronaut, have as many total daughters and as many Excellent daughters.

"As most of you probably know, Harborcrest Rose Milly, Astronaut's dam, was considered by many cowmen as the cow with as few faults as any cow has ever had. Her All-Time All-American status in 1965 I'm sure proved that.

"In the early 60's, Milly being a young, unproven cow with that high degree of perfection made her a difficult cow for me to evaluate as to her breeding needs. Since I am a believer in aAa methods of breeding evaluation, I listened carefully

Harborcrest Rose Milly (EX-97-3E-GMD), Astronaut's dam, was a star producer before he was born, receiving All-American honors and making 1242 lbs. fat.

to Bill Week's evaluation of the cow. After being unable to satisfy myself with finding a sire that would improve two traits of her, I settled on a sire that would definitely improve one. She was bred to our own herd sire at that time, Thonyma Ormsby Senator. She had a son by that mating and, as it was at the start of the outer space age, we called him Astronaut. Astronaut was a tall gangling calf and not exceptionally strong, which kept him from smoothing up.

"About that same time, Billy King stopped in at the farm while selecting consignments for the Maryland Convention Sale. Billy was delighted when he was successful in talking me out of the son of Milly.

"A few months later, at the convention in June, it is needless to say I was the most disappointed person there when Paclamar Astronaut sold for only $9000 to a syndicate of Maryland breeders. It was only a few months later that Curtiss Breeding Service leased Astronaut and his climb to fame began."

Astronaut's maternal side has been widely publicized — his dam the All-American 1242-lb. fat producer and dam of other famous offspring, Harborcrest Rose Milly (3E-97-GMD); her dam, Supreme Fay Marilyn (2E-91-GMD), a great brood cow in her own right. Milly's sire was Rainbow Sir Rose (VG), a minus sire.

Astronaut's $9000 sale price at the 1964 Convention Sale disappointed his owner, Dick Brooks of Paclamar Farm.

Many a discussion have centered around Astronaut's numbers. Geneticists cite him as an exception, traditionalists use him to support their "eye of the master" viewpoints, but in any case, he was the son of a minus bull.

Because Astronaut's sire, Thonyma Ormsby Senator (EX-94-TQ) had developed a rather sizable minus for production, his side of the pedigree has been disregarded, far too much according to some. Senator was much greater as a sire of type than of production. His 1436 classified daughters average 83.0 points actual and a high PDT of +2.70. At the same time, he has many high record daughters for production, although not enough to bring his average up.

Senator's sire, himself a minus bull, was justly famed Ormsby Burke Pontiac Mose (EX-92-GM), the Elmer Dawdy bred son of Wis Burke Ideal (VG-88-GM) and Posch Ormsby Johanna Bess (EX-92), dam of two high type, high-producing daughters.

Senator's dam was also recognized as one of the greats of her day. She was Thonyma Maudlene Senatora (2E), who made Gold Medal in 1960. A consistent producer and reproducer, she ended up with over 144,000 lbs. of 3.9% milk lifetime. Further, Senatora was by a strong breeding bull, Maytag Ormsby Fobes 14th (GP-82-GM), from the heart of the Princess family at Ed Reed's Thonyma Farm in Kansas. At the time Dick Brooks purchased the majority of the Thonyma herd in 1963, there were seven direct generations of the Princess family in the herd, all but one living and included in the purchase.

Breeding of this strength has to surface somewhere, and it surfaced with Paclamar Astronaut.

Ted Krueger, with Curtiss Breeding Service at the time, had serious doubts about Astronaut when he saw the bull's first milking daughter. He later said, "She resembled 'Marvelous Milly' only because both had four legs and a tail! The first Astronaut to freshen, in plain words, was at best a low Good Plus 2-year-old without much prospect of rising in the world.

"If they're all going to be like this, I told myself on the way back to the car, Astronaut is as good as dead. Maybe I am too — for helping get him to the stud."

But soon, Krueger's fears were calmed as more daughters went into production. It wasn't long before the Astronaut pattern started to emerge and

Dreamstreet Astro Buni (EX-96-2E) was Astronaut's highest scoring daughter.

make itself felt. Ted Harrison of New York came up with the Angie cow. She was one of the first, if not the very first, Astronaut daughter to go Excellent.

Word began getting around that Astronaut daughters made milk like they invented it. Angie gave Holstein breeders a preview of that blend of strength and dairyness, that inimitable style and that tremendous udder quality that were to make Astronaut famous as a sire of type. Angie was no fluke, either. In May, 1969, Astronaut came up with predicted differences of +1014 lbs. milk and +31 lbs. fat and +1.80 in type.

Astronaut's siring pattern seemed to stem from combining the style of his sire, Thonyma Ormsby Senator, with the dairyness of his dam, Harborcrest Rose Milly.

Astronaut has stood the test of time. In the process he has vindicated the faith that Billy King and friends placed in the rough-rumped son of a minus-proven bull. That same faith kept them from pushing the panic button when the first Astronaut daughter turned out to be an exception.

From his somewhat shaky start as a calf, Astronaut classified Very Good-85 at 2 years, going to 87 points at 4 years and to his present 90 points when mature. Always a plus bull for production and type, he made his Gold Medal in 1969.

For sheer number of offspring production-tested and classified, Astronaut trails only Round Oak Rag Apple Elevation. The July, 1984, USDA Sire Summary shows him with 49,858 tested daughters in 11,805 herds. Astronaut ranks second among all bulls in the breed with 34,885 classified daughters that average 82.0 age adjusted scores, giving him a PDT +1.57.

Astronaut's highest fat record daughter to date is Millerhurst Astro Dash-Twin (EX-92), owned by Rothrock Farms, Pennsylvania, who has 4-5 365d 3X 39,210 5.5% 2144. Astronaut has 1530 Excellent daughters, surpassed only by Round Oak Rag Apple Elevation.

Over the years, Astronaut has sired no less than 32 show animals good enough to be nominated for All-American by the *Holstein World*. Seven of the 32 became All-Americans, 10 became Reserve All-Americans and 13 received Honorable Mention. In 1983, the All-American 5-Year-Old, 4-Year-Old and Sr. 2-Year-Old all had Astronaut dams.

Astronaut appeared on the cover of *Wall Street Journal* in May, 1976 and made the AP and UPI wire services. His mother was billed as the greatest cow who ever lived. The million-dollar superstud left thousands of cattle behind him as proof. Astronaut's sons and daughters and their offspring number some 900,000. Today, Astronaut ranks second in the breed for number of tested daughters (49,858) and second for number of Excellent daughters (1530), second for Gold Medal daughters (89) and second for number of daughters over 100,000 lifetime (1502) ... not bad for a $9000 investment in 1964.

Round Oak Rag Apple Elevation

Ronald A. Hope and George A. Miller undoubtedly knew more about Round Oak Rag Apple Elevation's background and were closer to Elevation throughout his lifetime than any other individuals.

Ronald Hope is the owner and operator of Round Oak Farm with his son, Ronald Hope, Jr. A visit with Ronald Hope was a treat one would not soon forget. He could point out a cow and in a flash skip back three or four generations on the maternal side and tell you what kind of cows they were. In the next instant he would look forward two generations and tell you the matings he planned to make. Such abilities separate out the master breeders of Holsteins.

George Miller, former manager of Virginia Artificial Breeding Association and now director of marketing for Select Sires, spent many of his first 22 years on the farm of his uncle and cousin, Charles R. Hope and Ronald A. Hope, working with the herd. As a result he and Ronald became like brothers.

Starting with the first cow found in three-generation pedigrees of Elevation, his granddam, Round Oak Nettie Emaline (GP-84), Ronald Hope explains that the sire of Emaline, Glenafton Gaiety (EX-90-GM), was brought into the Round Oak herd thinking he could transmit size. However, when checking the Gaietys in the calf and heifer lots, the Hopes determined that Gaiety was not doing as good a job as they had hoped.

Looking for a bull that would sire the kind of stature and upstandingness they wanted, Hope listened when George Miller, then manager of Virginia ABA, told them about a great, big, tall bull named Osborndale Ivanhoe. Hope went up to

Round Oak Rag Apple Elevation 1491007
Born: 8-30-65 Excellent-96 Gold Medal 7/84

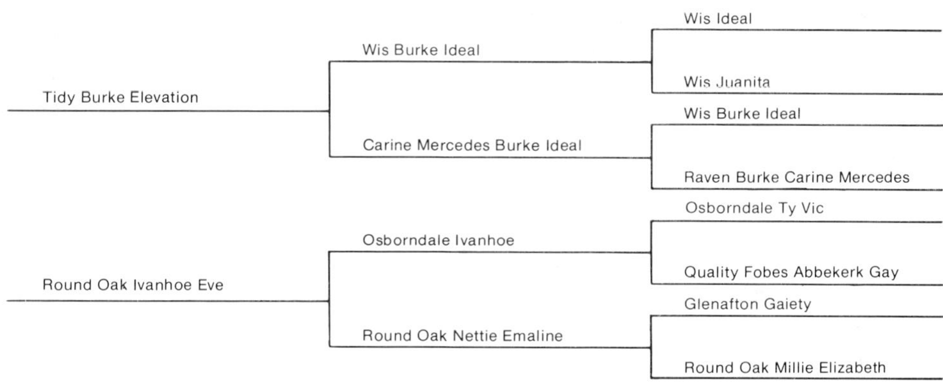

Atlantic Breeders Cooperative to take a look at Ivanhoe and was so impressed with his massive frame that he bought 50 ampules on the spot for $2.50 each. Ivanhoe semen was used to breed Emaline producing Round Oak Ivanhoe Eve (4E-94).

As a heifer, Eve was much like the other Ivanhoes in the herd, a stretchy, gangly heifer that took some time to mature. Eve scored VG-85 as a 2-year-old and made 13,260 milk and 562 fat in 311 days after calving at 2y1m of age.

Round Oak Ivanhoe Eve (EX-94-4E) was Elevation's dam.

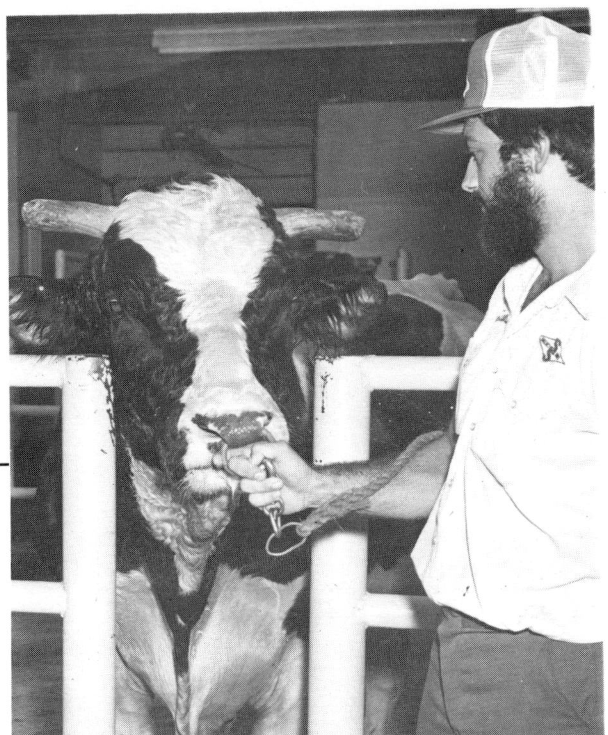

Elevation, shown with handler Steve Raber, at 13 years.

The young milking Ivanhoe daughters in the herd looked like they needed strength. Once again George Miller had a suggestion — Tidy Burke Elevation. However, this time he knew his idea would not be as easily sold, since the Hopes had always liked following Rag Apple breeding. To convince Ronald Hope to use this inbred Burke bull (a Wis Burke Ideal son out of a Wis Burke Ideal daughter), Miller looked up information on Tidy Burke Elevation and had graphs made showing how the daughters ranked well above breed average in terms of strength and shape of udder. Thus, the decision was finally made to breed Eve and a few other Ivanhoe daughters to Tidy Burke Elevation. The resulting calf, born on August 30, 1965, was Round Oak Rag Apple Elevation.

An in-depth study of Round Oak Rag Apple Elevation's pedigree demonstrates how, before the bank of present-day genetic knowledge, the great breeders of that day developed their breeding theories and systems, and put them into practice ever so carefully. On his maternal side, Elevation traces no less than 20 times to Johanna Rag Apple Pabst, founder of the Rag Apple bloodlines.

On his paternal side, he is intensely Burke, tracing 11 times to Wisconsin Admiral Burke Lad, founder of the Burke bloodlines. Elevation is therefore a beautiful blend of two bloodlines. One may consider Elevation to have an old-timer's "dream pedigree," and in this case, that dream has come true.

Elevation's July, 1984, USDA proof shows predicted difference of +366 Milk +22 Fat and +$61 Value on 54,843 AI daughters located in 10,280 herds. These 54,843 AI daughters averaged, on an ME basis, 18,440 lbs. milk 3.7% 673 lbs. fat. This number of 54,843 tested daughters established a world record for any sire.

A memorial to Elevation was unveiled in 1979. From the left: Mr. and Mrs. Ronald Hope, Robert Rumler, Richard Chichester and George Miller of Select Sires.

One benchmark of success in milking cows is the 1000-lb. fat record. As of March 1, 1982, 1000 registered daughters of Elevation had produced 1000 lbs. fat or more on DHIA.

Another benchmark of success is production of 30,000 lbs. of milk in one lactation of 365 days or less. Two hundred Elevations have achieved this goal. The highest milk record daughter is Northcroft Ella Elevation (EX-97-3E), All-Time All-American Aged Cow. At 7-7 365d, Ella made 48,731 4.2% 2028. One of the highest fat record daughters is Braintrim Elevation Marikin (VG-88), owned by Lekker Holsteins, California. At 4-3 365d 3X, she made 40,650 5.2% 2104.

His July, 1984, HFAA type proof is +2.19 PDT on 40,961 classified daughters. These daughters had an average age adjusted classification score of 83.0 points. As an individual he was classified EX-96 at 10 years. His dam, Round Oak Ivanhoe Eve, was EX-94 at 9 years, EX-93 at 8 years, and EX-92 at 5 years.

As of November, 1984, Elevation has sired 2862 Excellent daughters. In June, 1976, when he was 10 years of age, Elevation had 30 Excellent offspring. However, in the next 101 months, he gained 2832 more Excellents, approximately 28 additional Excellents per month, or almost one a day.

"Big E" is the first sire in the breed to have +10,000 registered sons, which verifies his tremendous popularity with the total registered industry. Elevation has shown that he can transmit both type and production. He is the first sire in the breed to exceed 2000 Excellent daughters (2862), and to have more than 2000 daughters with records over 100,000M in their lifetime (2341).

Elevation progeny have 89 All-American nominations to date (1977-1983). He has sired 18 individual All-American winners plus the All-American Get of Sire in 1977, 1978, and 1983. But the most significant of all show ring awards is the All-Time All-American recognition sponsored by the *World*. There are four milking class 1984 All-Time All-Americans, three by Elevation and the fourth a maternal granddaughter. Ella (EX-97) is the All-Time All-American Aged Cow. Shadowcliff R A Gina (EX-97) is the All-Time All American 2-Year-Old, 3-Year-Old and Reserve 4-Year-Old. The All-Time All-American 4-Year-Old, Brookview Tony Charity (EX-97), was sired by a Matt son and out of an Elevation daughter. Elevation's 1983 All-American Get of Sire was Honorable

Mention All-Time All-American. As astounding as his daughters are, the sons of Elevation, who include Bova, Tradition, Mars, Warden, and a host of others, measure right up to them.

Glendell Arlinda Chief

Glendell Arlinda Chief became one of the prominent sires of the breed after a very inauspicious start. His pedigree was not particularly noteworthy because his dam died of unknown causes before finishing her first record; he had health problems as a calf and nearly died from bloat before leaving Glendell Farm; he was not very widely sampled and his first proof (PD68 +507M with 20% repeatability) did not impress the AI firms who reviewed it. In spite of this, Glendell is one of the Cinderella stories of the Holstein industry and is now well on his way to becoming one of the greatest milk bulls in Holstein history. He has earned his reputation by transmitting strength and milk production. In the process, he has proven that sometimes the instincts and "cow sense" of breeders are as important as genetics.

Glendell's sire was Pawnee Farm Arlinda Chief who is described earlier in this chapter. Chief had recently come to Curtiss Breeding Service and was just starting to gain his reputation for milk yield. (His first proof, with a PD of +1845 was issued in 1968, the same year Glendell was born.) The dam of Glendell, Glendell Ned Boy Adorn, was part of a cow family that Ken and JoAnn Hartman had developed. JoAnn Hartman explained, in a 1984 interview for the *Holstein World,* why they decided to keep Glendell as a calf after his dam had died unclassified and untested (195 days): "I guess you sometimes let feelings enter into some animals more than genetics ... we just thought that Glendell's dam was one of the top animals we had bred at that time, so we decided to use her son. We also liked this family because we had bred the dam, the granddam and the great granddam."

The Hartmans' instincts were right, but it would take time to find that out. They had not promoted Glendell as a young sire and AI firms wanted to "wait and see" after seeing his first proof.

"We really found Glendell Arlinda Chief by accident," explains Phil Jayne,

Glendell Arlinda Chief 1556373
Born: 11-19-68 Excellent-93 Gold Medal 7/84

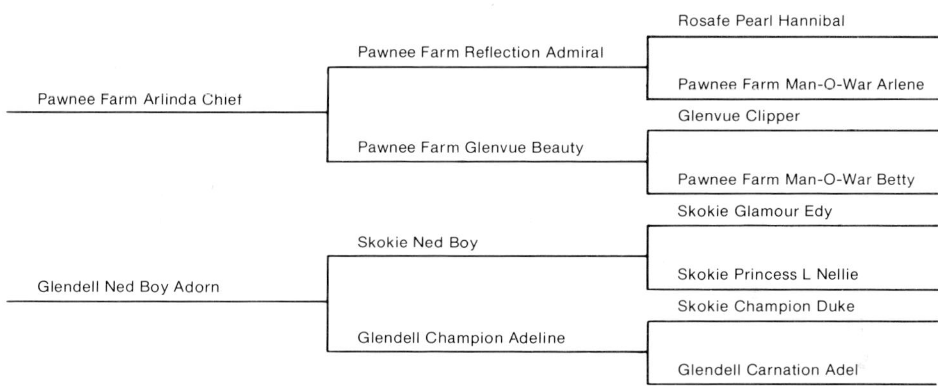

former sales coordinator for Select Sires and later an independent semen salesman in California. Jayne was traveling in southern Illinois with Prairie State/Select Sires manager Ray Hess in the fall of 1973 when an area technician mentioned that the Ken Hartman family had a bull that they should see. "Since we had some spare time, we decided to visit Glendell Farms," he recalls, "and after evaluating Glendell and all of his daughters we became quite excited. His daughters had plenty of size and substance and appeared to be the kind of cattle that could milk quite heavy and stay at it. I also felt that Chief would work well on a Ned Boy daughter."

But Jayne realized he had an uphill battle to get Glendell into the Select Sires' lineup: "With his low proof and the fact that his dam never completed her 2-year-old lactation, Glendell had two big strikes against him. I knew I had a real sales job when I returned to Select Sires. I guess it was more of a gut feeling than anything else, because I don't claim to have been smart enough to know that Glendell would ever be as popular as he is today. At the time Select Sires didn't have a proven son of Arlinda Chief, and I didn't feel like we could afford to let him go. There had already been some other organizations in to look at Glendell

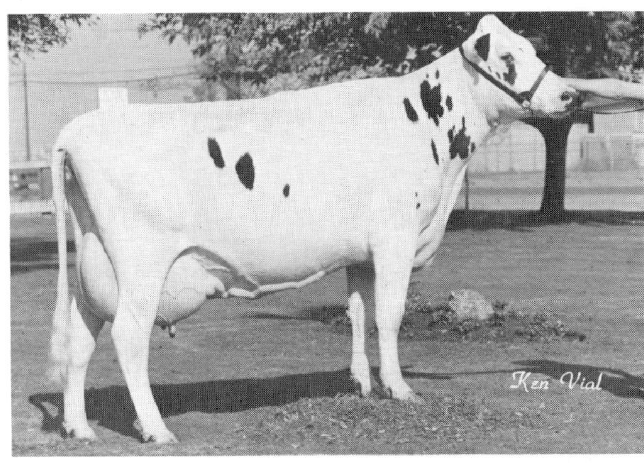

Janibo Chief Quinsy (VG-88) has a 365d record of 45,910 4.9% 2234.

and I thought if we didn't hurry, someone else was going to get him."

The AI firm, after further evaluation, brought Glendell into the stud and he soon began to earn his reputation of siring high-milking daughters.

Glendell's January '84 USDA production summary includes 20,498 daughters, a count that surpasses that of his sire, Arlinda Chief, with 17,250 daughters. His PD milk of +992 and +$120 keeps him among the breed's elite, even though it has been five years since he died.

His specialty is siring high index daughters. The January '84 cow indexes show Glendell to be the breed's leading sire of cows with over +$100 indexes with an average test of at least 3.0%. He sired 441 of the 3520 cows in this category — his top 50 daughters are +$145 or higher. Glendell daughters have completed 298 records over 30,000 lbs. milk with 54 of them over the 35,000-lb. mark. His offspring have completed 719 records over the 1000-lb. fat mark, with 68 of them over 1300 lbs.

Glendell's Predicted Difference for Type calculated by the Holstein Association has also held up well. The July '84 figure shows +1.23 with 12,747 daughters included. Glendell gets some fancy ones, too, as evidenced by the number of Excellent offspring — 292. Of them, five are EX-94, five are 93, 17 are 92 points.

Cor-Vel Enchantment is Glendell's highest son for PD Milk (+1928).

Ken and JoAnn Hartman, here shown with a statue of Glendell, had a hunch about Glendell's potential, a hunch that paid off for them and the U.S. Holstein breed.

Glendell continues to hold his own among the current elite sires of the breed as measured by his +582 TPI (Total Performance Index).

Glendell has also gained notice as a sire of high production sons, just as he has as a sire of high index daughters. On that same Top 100 TPI List, there are 11 sons of Glendell. Cor-Vel Enchantment ranks first among them for TPI (third among all bulls) and PD Milk. Aufdermaur Glendell Lincoln is the top Glendell son for PDT.

Since his offspring boast these top credentials, it comes as no surprise that they have commanded top dollar in the auction sale ring. There have been 16 Glendell daughters sold for $50,000 or more at auction, another 28 have brought over $25,000, and six sons have been sold at auction for at least that price. In 1983, there were 27 offspring of Glendell sold for $10,000 or more, including the then world record price Holstein male, Woodbine Gold Duster-ET, struck off at $600,000.

Glendell was much like his sire, Pawnee Farm Arlinda Chief. He transmits type and production. Glendell sired 11 of the 100 top bulls on the 1984 Total Performance Index listing by the Holstein Association. He leads the breed by having nine daughters over 40,000 milk (their average classification is VG-85). He is second in the breed for +PD milk for sires with 100 or more proven sons. Yet, on some farms, he never would have had the chance.

JoAnn Hartman put it like this, "Don't overlook a sire just because his dam died early. He could be another Glendell."

Appendix

PAST PRESIDENTS OF THE HOLSTEIN ASSOCIATION

C. R. Payne
New York
1896-97

W. A. Matteson
New York
1898-1901

W. J. Gillett
Wisconsin
1901-03

Henry Stevens
New York
1903-04

A. A. Cortelyou
New Jersey
1904-06

O. U. Kellogg
New York
1906-09

Charles W. Wood
Massachusetts
1909-12

A. A. Hartshorn
New York
1912-14

D. D. Aitken
Michigan
1914-21

Frank O. Lowden
Illinois
1921-30

Prof. Henry H. Wing
New York
1930-31

Dr. L. M. Thompson
Pennsylvania
1931-32

A. J. Glover
Wisconsin
1933-37

A. C. Oosterhuis
Wisconsin
1937-42

W. S. Moscrip
Minnesota
1942-47

Carl G. Wooster
New York
1947-48

Owen D. Young
New York
1948-49

S. B. Hall
Oregon
1949-51

Harold J. Shaw
Maine
1951-53

J. Homer Remsberg
Maryland
1953-55

Albert B. Craig
Pennsylvania
1955-57

Scott Meyer
Missouri
1957-59

Leon A. Piguet
New York
1959-61

Fred J. Nutter
Maine
1961-63

Russell L. Pfeiffer
Washington
1963-65

A. J. Johnson
Iowa
1965-67

R. DeWitt Mallary
Vermont
1967-69

Harold W. Craun
Virginia
1969-71

A. C. Thomson
Illinois
1971-73

Clifford E. Bailey
Washington
1973-75

Gordon W. Newton
South Carolina
1975-77

W. R. (Dick) Brooks
Colorado
1977-79

Ivan K. Strickler
Kansas
1979-81

Maurice R. Keene
Maine
1981-83

Dr. David G. Smokler
Texas
1983-85

HOLSTEIN ASSOCIATION
75 YEARS OF FINANCIAL GROWTH

YEAR ENDED	INCOME	EXPENSE	TOTAL FUND BALANCES	TOTAL INVESTMENTS	OTHER NET ASSETS
MAY 15, 1907	42,000	30,000	52,000	29,000	23,000
MAY 15, 1912	106,000	98,000	145,000	108,000	37,000
APR. 30, 1917	212,000	197,000	334,000	247,000	87,000
DEC. 31, 1922	391,000	394,000	466,000	389,000	77,000
DEC. 31, 1927	406,000	442,000	406,000	327,000	79,000
DEC. 31, 1932	185,000	202,000	301,000	171,000	130,000
DEC. 31, 1937	223,000	205,000	254,000	101,000	153,000
DEC. 31, 1942	371,000	295,000	496,000	222,000	274,000
DEC. 31, 1947	685,000	697,000	889,000	637,000	252,000
DEC. 31, 1952	1,055,000	1,039,000	1,292,000	707,000	585,000
DEC. 31, 1957	1,210,000	1,198,000	1,528,000	1,073,000	455,000
DEC. 31, 1962	1,607,000	1,599,000	1,882,000	1,319,000	563,000
DEC. 31, 1967	2,065,000	1,997,000	1,936,000	1,600,000	336,000
DEC. 31, 1972	3,147,000	2,661,000	3,065,000	2,764,000	301,000
DEC. 31, 1977	4,835,000	4,485,000	4,399,000	4,347,000	52,000
DEC. 31, 1982	10,520,000	10,144,000	6,829,000	5,964,000	865,000

HOLSTEIN REGISTRATIONS - YEAR BY YEAR

Year	Females	Males	Total
1885-6	3,667	2,314	5,981
1886-7	3,709	3,109	6,818
1887-8	4,959	4,225	9,184
1888-9	4,392	3,470	7,862
1889-90	4,323	2,617	6,940
1890-1	5,096	2,197	7,293
1891-2	4,462	1,759	6,221
1892-3	3,466	1,212	4,678
1893-4	2,800	1,049	3,849
1894-5	2,348	839	3,187
1895-6	2,297	773	3,070
1896-7	1,885	651	2,536
1897-8	2,225	738	2,963
1898-9	4,695	1,784	6,479
1899-00	3,381	1,365	4,746
1900-01	3,648	1,460	5,108
1901-02	4,252	1,738	5,990
1902-03	4,753	2,088	6,841
1903-04	5,567	2,477	8,044
1904-05	6,547	3,226	9,773
1905-06	7,918	3,842	11,760
1906-07	9,809	4,841	14,650
1907-08	10,850	5,684	16,534
1908-09	12,570	7,021	19,591
1909-10	16,487	9,689	26,176
1910-11	20,417	12,472	32,889
1911-12	23,792	13,743	37,535
1912-13	26,951	16,364	43,315
1913-14	29,750	18,336	48,086
1914-15	42,063	25,617	67,680
1915-16	46,549	26,116	72,665
1916-17	49,098	24,749	73,847
1917-18	59,549	28,730	88,279
1918-19	60,589	30,298	90,887
1919-20	77,712	36,791	114,503
1920-21	88,265	39,585	127,850
*1921	46,590	19,218	65,808
1922	83,140	30,631	113,771
1923	86,043	29,089	115,132
1924	83,320	28,209	111,529
1925	82,659	26,935	109,594
1926	82,971	28,117	111,088
1927	81,146	28,817	109,963
1928	88,214	33,512	121,726
1929	89,927	35,438	125,365
1930	75,901	29,242	105,143
1931	70,535	21,811	92,346
1932	54,481	13,834	68,315
1933	83,012	15,521	98,533
1934	82,935	17,283	100,218
1935	61,665	15,220	76,885
1936	61,154	16,788	77,942

Holstein Registrations *(Continued)*

Year	Females	Males	Total
1937	62,421	16,689	79,110
1938	64,027	17,595	81,622
1939	67,043	18,555	85,598
1940	119,482	25,941	145,423
1941	78,985	23,818	102,803
1942	80,575	26,049	106,624
1943	82,563	28,634	111,197
1944	93,917	28,993	122,910
1945	88,821	24,625	113,446
1946	130,818	38,520	169,338
1947	119,655	33,084	152,739
1948	132,762	35,576	168,338
1949	141,596	36,329	177,925
1950	147,584	36,662	184,246
1951	155,009	36,629	191,638
1952	157,266	32,424	189,690
1953	163,810	25,752	189,562
1954	172,424	23,539	195,963
1955	174,139	23,469	197,608
1956	182,682	22,845	205,527
1957	189,923	22,522	212,445
1958	228,831	26,036	254,867
1959	245,818	28,095	273,913
1960	241,878	23,983	265,861
1961	242,125	21,489	263,614
1962	244,427	20,674	265,101
1963	239,820	18,748	258,568
1964	243,864	18,555	262,419
1965	238,713	18,056	256,769
1966	251,483	18,894	270,377
1967	256,146	20,729	276,875
1968	271,423	20,705	292,128
1969	243,457	17,687	261,144
1970	261,250	20,324	281,574
1971	286,576	22,913	309,489
1972	256,299	21,552	277,851
1973	262,053	23,766	285,819
1974	268,331	23,458	291,789
1975	258,838	20,308	279,146
1976	274,027	21,856	295,883
*1977	301,521	29,094	330,615
1978	278,606	26,883	305,489
1979	301,590	29,101	330,691
1980	322,801	31,148	353,949
1981	343,793	33,173	376,966
1982	352,851	34,047	386,898
1983	387,951	37,434	425,385

*Eight months

*1977-1983 — Since 1977, male/female registration breakdowns have not been listed separately. The breakdown figures shown are estimated from average percentages.

TOP 50 LIFETIME MILK LEADERS

Cow / Owner	Milk	Fat	%
Breezewood Patsy Bar Pontiac 6174402 Gelbke Brothers, OH	425769 Milk	19203 Fat	4.5%
Or-Win Masterpiece Riva 4624699 Willard Behm & Son, MI	346218	11329	3.3
Maplewood Lane Dora Mega 4448454 Jim N. Beane, IL	337840	11678	3.5
College Ormsby Burke 3420439 Colorado State Univ., CO	337292	11351	3.4
Orchard Meadows Perfection 4645828 O. J. Thrall, Inc., CT	329400	12124	3.7
Kuehn-Acres Pioneer Duchess 6562751 Joseph Kuehn, WI	325984	12172	3.7
Skagvale Graceful Hattie 6069325 Tenneson Brothers, WA	319710	11218	3.5
Triumph Pietje Ormsby Ann 5453892 Leonard Kacuba, PA	318243	13031	4.1
Artevel Carnation Gem 6331996 Artevel Farms, R. A. Tevelde, CA	316080	10587	3.3
Green Hilltop Duchess Ann 4977892 Robert K. Reed, VA	312448	9619	3.1
Maiden-Valea Fond Apple Lad 7178479 Marlu Farm, NJ	309590	9901	3.2
Nor-Lene Alstar Ethel 7205064 Norman Lee Becher, IN	309163	12654	4.1
Schoch Lucky Lady-Twin 9479886 Schoch Brothers, CA	308753	11370	3.7
Zeldenrust Pontiac Korndyke 2649438 Ray Bottema, IN	306051	11649	3.8
Betterway Ideal Neptune Maid 4719419 Joe Pimentel, CA	305398	12290	4.0
Lakefield Fobes Delight 3232651 Carnation Milk Farms, WA	304064	10987	3.6
Olmar Polly Bob Faithfull 6956834 Roger & Floyd Marti, MN	300340	10021	3.3
Korndyke Beets Jannek Segis 2065418 Clark Bowen & Son, PA	299459	10748	3.6
Rag Apple Maestro Ormsby 5193132 Gerrit J. Buth & Sons, MI	295490	10601	3.6
Mal-Rob-Ric Nettie Ken 6504332 Gilbert P. Robery, MA	292710	10053	3.4
Kay Hart Jingle Ormsby Gettie 4491935 Anton Betschart, CA	288033	10901	3.8
Pleasantland Magician Jeanette 6632841 Richard A. Muller, IL	285930	9799	3.4
Verboon Randy Eunice 6763304 Paulo Brothers, CA	284450	10642	3.7
Morrisville Deen Seeley 4143013 St. Univ. New York, NY	283290	10099	3.6
Pride Skyliner Kathy 5889338 Curti Farms, CA	283173	9674	3.4

Top 50 Lifetime Milk Leaders *(Continued)*

Walbea Kathy Triune 6800321 Terry Blickensderfer, OH	282871	9344	3.3
Kanza Skyhawk Tona 5165522 Paclamar Farms, CO	282810	9418	3.3
St Croixco Ideal Jess Rose 5242411 St. Croix County Hospital, WI	282617	11851	4.2
Blossomelle Ivanhoe Jerry 6366596 Rhelda E. Royer, PA	282610	10790	3.8
Minnow Creek Eden Delight 2494802 Meadow Farm Dairy, VA	282278	12211	4.3
C Howcroft Star Bright 7547218 Russell & Maynard Axelson, WA	281740	10733	3.8
Wildmoor Magic Lass-Twin 5545671 Lynn & Bonnie Miller, PA	281462	9638	3.4
Pansco Hazel 1817843 F. F. Pellissier, CA	281193	10599	3.8
Vernway Echo Violet 6859138 Paclamar Farms, CO	280590	8082	2.9
Willow Maple Reflection Mae 6594935 Willow Maple Farm, PA	280196	10172	3.6
Howacres Posch Joan 2875511 Gardner L. Lewis, Jr., VT	280017	9303	3.3
Lantland Fond Hope Daisy 6777196 Fred Lant, NY	279850	10160	3.6
Nunesdale Noble Lyndia 5951364 Larry A. Shehadey, CA	279060	9745	3.5
Lantland Ormsby Ideal 4448263 Fred Lant, NY	278959	11055	4.0
Lantland Ormsby Marcha 6860285 Fred Lant, NY	278660	8184	2.9
Coxland Comet Ideal Lady 6337257 Muriel E. Steubing, TX	277280	8871	3.2
Pleasantland Mabel Roxette 5176840 Ralph A. Muller & Sons, IL	276435	8578	3.1
Artevel Brigitte Lea 6746844 Marvin J. Tevelde, CO	275970	9287	3.4
Belleview Pioneer Octane 6973197 Donald & Keith Long, WI	275930	10125	3.7
Robthom Blondie Voyageur 5189176 Robert F. Thomson, Jr., MO	275634	9440	3.4
Fultonway Ivanhoe Rae 4977152 J. Mowery Frey & Son, PA	275553	12716	4.6
Pellholm Black Eagle Dolly 8489976 Frederick Barringer, NY	275296	9198	3.3
Ocean-View Baroness Mimi 6771278 Marvin L. Nunes, CA	275240	10793	3.9
Kanzabrook Matt Missy 7274432 Jackson Holsteins, CO	275040	10328	3.8
U N H Dauntless Nova 2874989 Univ. of New Hampshire, NH	274770	10109	3.7

AVERAGE PD82 OF U.S. HOLSTEIN SIRES OF ANIMALS BORN SINCE 1960

Year of Birth	Males PDMilk (lbs)	Males PDFat (lbs)	Females PDMilk (lbs)	Females PDFat (lbs)
60	-1089.0	-32.1	-1058.8	-29.7
61	-1090.0	-32.0	-1064.2	-30.0
62	-1132.0	-33.9	-1088.1	-31.1
63	-1135.0	-34.3	-1067.8	-31.4
64	-1138.6	-34.4	-1065.6	-31.6
65	-1127.7	-33.6	-1039.3	-30.6
66	-1100.8	-32.0	-1003.5	-28.7
67	-1109.6	-31.3	-1003.2	-28.6
68	-1083.7	-29.6	- 988.9	-28.0
69	- 984.2	-27.0	- 922.9	-26.9
70	- 924.0	-27.1	- 880.3	-26.7
71	- 798.4	-24.1	- 814.3	-25.2
72	- 640.0	-17.8	- 713.3	-21.7
73	- 535.4	-14.5	- 641.0	-19.8
74	- 490.9	-13.9	- 596.7	-18.6
75	- 415.0	-12.7	- 542.6	-17.0
76	- 258.4	- 8.0	- 398.9	-12.9
77	- 106.4	- 2.7	- 256.8	- 8.5
78	.8	.5	- 145.3	- 5.5
79	98.6	4.4	- 31.5	- .7
80	239.4	9.8	80.4	4.2
81	412.2	15.7	220.9	9.6
82	554.4	21.5	361.7	14.5
83[a]	748.5	29.4	539.9	20.9
84[a]	895.7	35.6	687.1	26.0
85[a]	1043.0	41.7	834.4	31.2

[a] All values are projected estimates

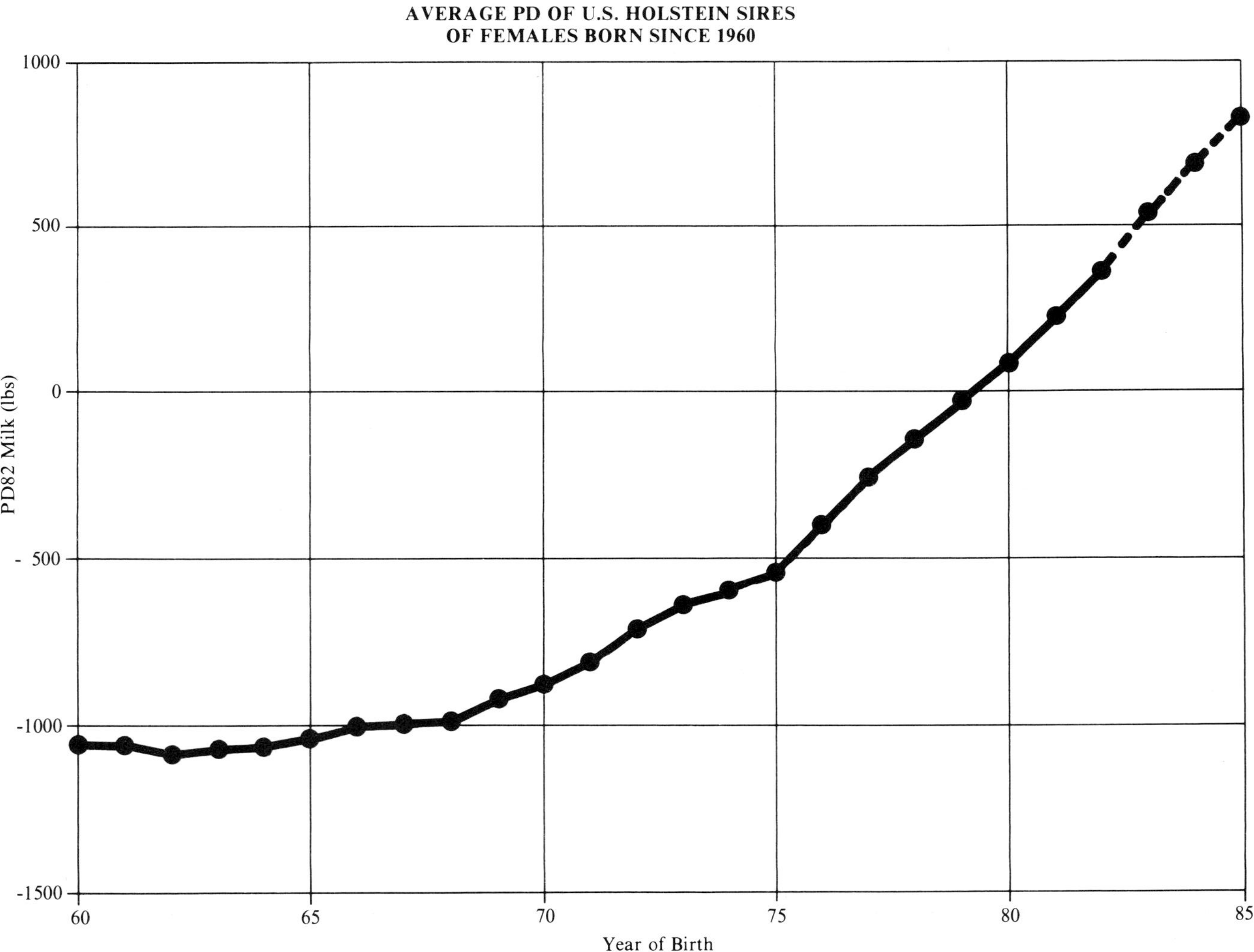
AVERAGE PD OF U.S. HOLSTEIN SIRES
OF FEMALES BORN SINCE 1960
PD82 Milk (lbs)
1000
500
0
- 500
-1000
-1500
60
65
70
75
80
85
Year of Birth

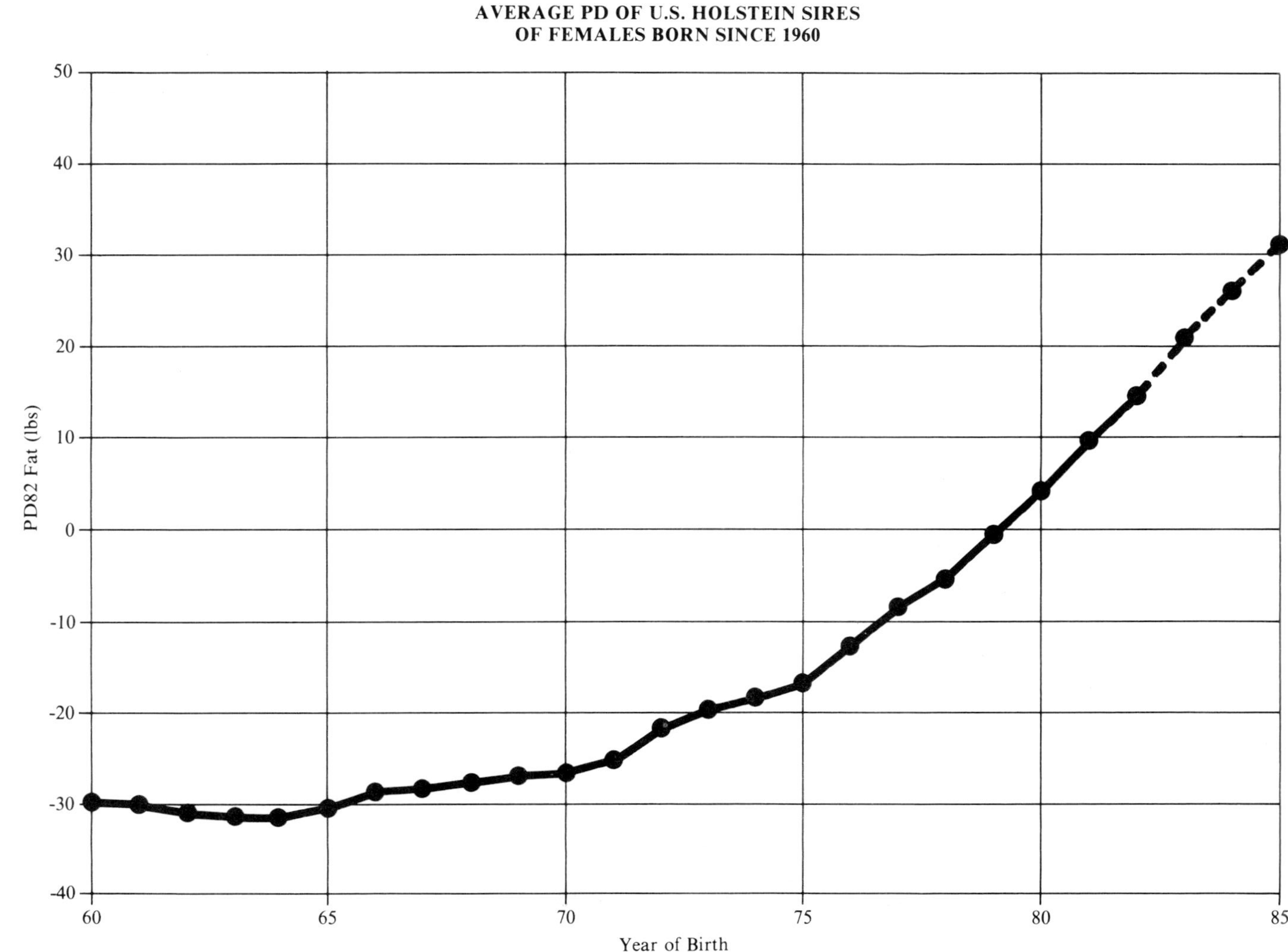
AVERAGE PD OF U.S. HOLSTEIN SIRES
OF FEMALES BORN SINCE 1960
PD82 Fat (lbs)
50
40
30
20
10
0
-10
-20
-30
-40
60
65
70
75
80
85
Year of Birth

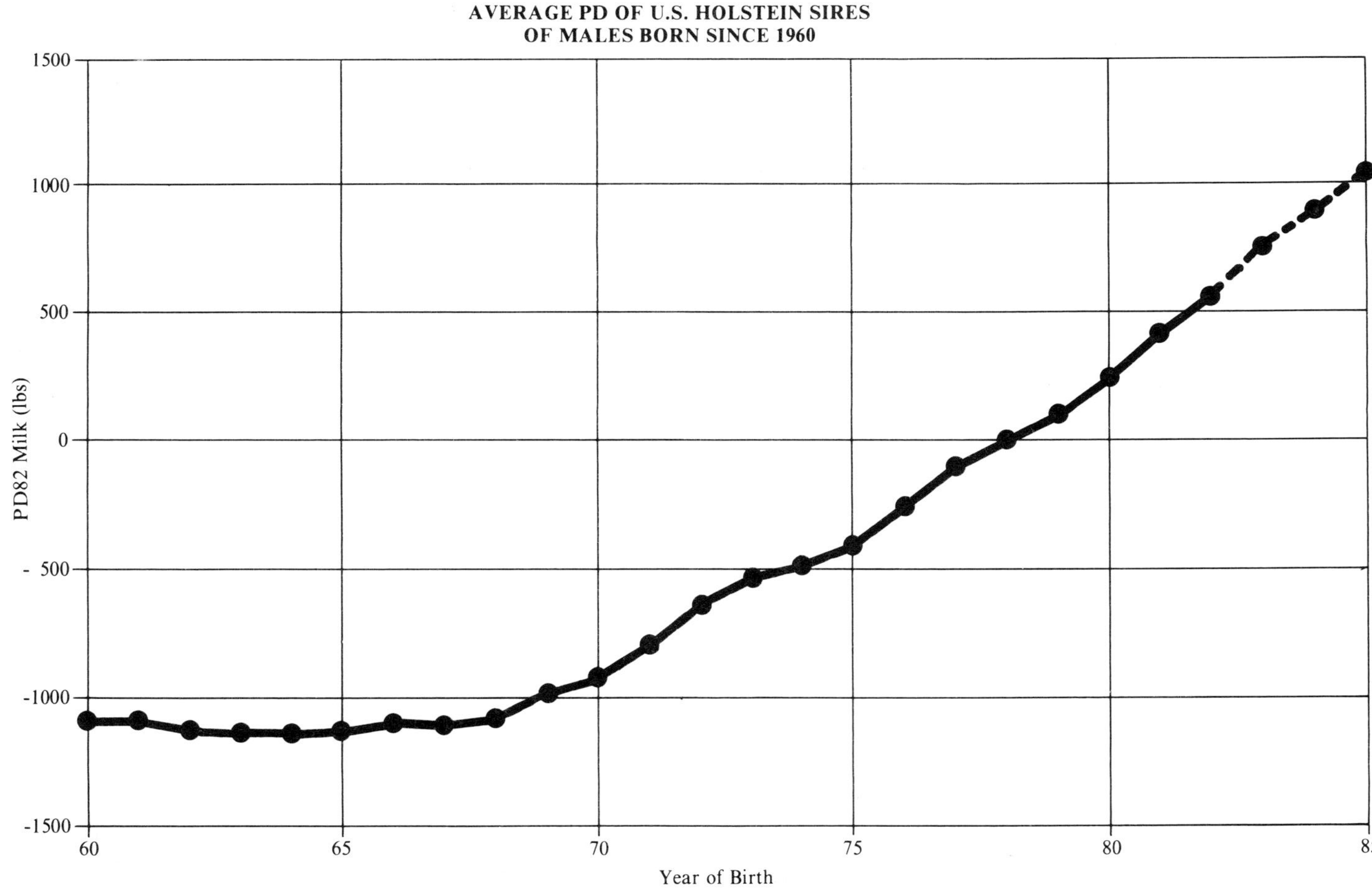
AVERAGE PD OF U.S. HOLSTEIN SIRES
OF MALES BORN SINCE 1960
1500
1000
500
0
- 500
-1000
-1500
PD82 Milk (lbs)
60
65
70
75
80
85
Year of Birth

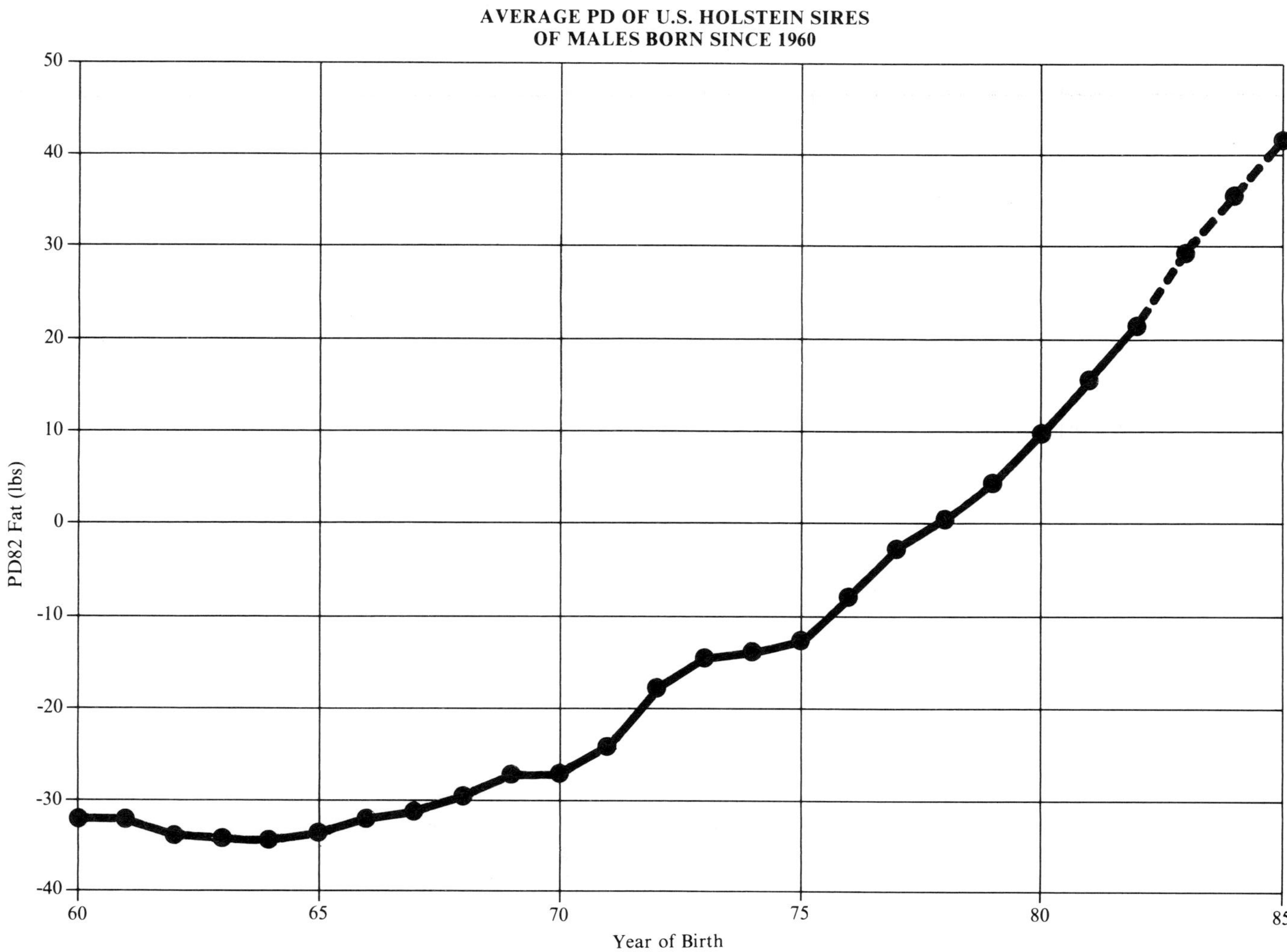
AVERAGE PD OF U.S. HOLSTEIN SIRES
OF MALES BORN SINCE 1960
PD82 Fat (lbs)
50
40
30
20
10
0
-10
-20
-30
-40
60
65
70
75
80
85
Year of Birth

NATIONAL CONVENTION SALES 1904-1932

Year	Place	Sale Manager	No. Head	Total	Ave.	Top	
1904	Syracuse, NY	S. D. W. Cleveland	160+	$31,025	$183.00	$1,200F	Pietje 22d (imported)
1905	Syracuse, NY	S. D. W. Cleveland	135		140.00	1,000F	Pontiac Netherland De Kol
1906	Syracuse, NY	S. D. W. Cleveland	159	36,895	232.00	2,600F	Aaggie Tehee De Kol
1907	Syracuse, NY	S. D. W. Cleveland	155	39,950	258.00	1,250F	Pontiac Jewel
1908	Syracuse, NY	S. D. W. Cleveland	156	36,200	232.00	1,300F	Elzevir Barnum De Kol
1909	Syracuse, NY	S. D. W. Cleveland	214	66,720	310.00	1,800F	Bloomingdale Hengerveld Aaggie
1910	Syracuse, NY	Cleveland Sales Co.	171	61,830	361.00	2,600F	Pauline Jamaica De Kol
1911	Syracuse, NY	Liverpool Sales & Ped. Co.	218	92,115	422.00	10,000M	King Segis Pontiac Alcartra
1912	Syracuse, NY	Liverpool Sales & Ped. Co.	229	74,500	325.00	3,000M	King Segis Pontiac Manor
1913	Syracuse, NY	E. M. Hastings	197	95,386	475.00	3,000F	Bloomingdale Hengerveld Palmyra
1914	Chicago, IL	T. E. Getzelman	179	107,855	603.00	20,000M	King Segis Pontiac Chicago
1915	Cazenovia, NY	Liverpool Sales & Ped. Co.	202	88,865	441.00	3,650F	Fairview R A De Kol
1916	Detroit, MI	H. A. Moyer	143	156,710	1096.00	20,000M	King Champion R A
1917	Worcester, MA	H. A. Moyer	143	296,570	2,073.00	53,200M	King O Jane R A
1918	Milwaukee, WI	H. A. Moyer	175	425,810	2,433.00	106,000M	Carn. King Sylvia
1919	Philadelphia, PA	H. A. Moyer	200	342,270	1,712.00	26,000F	Rolo Mercena De Kol
1920	St. Paul, MN	MN H-F Assn.	237	728,775	3,075.00	50,000M	Alcartra King Sylvia
1921	Syracuse, NY	NY H-F Assn.	172	138,445	805.00	4,250F	Arcady Segis Atia 2d
1922	Kansas City, MO	W. H. Mott (MO Hols. Br.)	144	69,445	469.00	3,200F	Wadmantje Finderne Pride Pontiac
1923	Cleveland, OH	Ohio Holstein Assn.	115	69,000	600.00	4,100M	Avon Pontiac Echo Lad
1924	Richmond, VA	W. L. Kirby (VA Hols. Br.)	148	38,200	260.00	1,000F	Star Segis Pontiac Superior
1925	Grand Rapids, MI	MI Holstein Assn.	58	44,250	763.00	3,700M	King Rose Ormsby
1926	Des Moines, IA	Iowa Holstein Br. Assn.	54	40,350	747.00	3,500F	Miss Mutual Rose De Kol
1927	Springfield, MA	New England H-F Assn.	70	28,025	400.00	1,225F	Cattaraugus Flora Colantha
1928	Milwaukee, WI	Wisconsin H-F Assn.	74	41,195	556.69	4,500M	Dutchland Denver The Great
1929	Philadelphia, PA	Paul B. Misner	141	94,568	670.70	10,100M	Baynewood Calamity Ormsby
1930	Denver, CO	W. L. Baird	85	33,710	396.58	6,200M	Carnation Inka Matador
1931	Syracuse, NY	D. W. McLaury	124	43,625	351.73	2,050F	Springbank Countess Finderne
1932	Madison, WI	Wisconsin H-F Assn.	60	7,660	127.67	600M	Carnation Inka Prince

NATIONAL CONVENTION SALES 1933-1961

Year	Place	Sale Manager	No. Head	Total	Ave.	Top	
1933	Chicago, IL	Melin-Petersen Co.	133	32,155	242.00	1,750F	Ellenvale Bonheur O Posch
1934	Ellicott City, MD	Paul B. Misner	60	21,950	365.83	2,450M	Carnation Governor Prospect
1935	(Conv. at Seattle - no sale with it)						
1936	Indianapolis, IN	Melin-Petersen Co.	89	22,345	251.00	1,600M	Man-O-War Concentrate
1937	Deerfield, IL	Baird & Darcey	86	33,907	394.27	2,200F	Bessie Ormsby Fobes Maid
1938	Oconomowoc, WI	Paul B. Misner & A. W. Petersen	88	32,560	370.00	1,250F	Oostie Inkarnation Bussie
1939	Earlville, NY	R. Austin Backus	85	21,827	256.00	1,625M	Carnation Forecaster
1940	Oconomowoc, WI	Paul B. Misner	93	32,490	349.35	2,150F	Femco Lady Pride Piebe Butter King
1941	East Lansing, MI	Paul B. Misner	98	34,115	348.11	3,200M	Carnation King
1942	Minneapolis, MN	Paul B. Misner	90	42,055	467.28	2,500M	Carnation Perfection Triumph
1943	No convention						
1944	Columbus, OH	Paul B. Misner	79	109,980	1,392.15	20,000F	Montvic Bonheur Pietje B
1945	No convention						
1946	Cary, IL	Baird & Darcey	50	154,125	3,082.50	23,000M	Bancroft Acres 23 Grand
1947	Springfield, MA	New England H-F Assn.	71	77,450	1,090.85	4,300F	Tyler Farms R A May
1948	Kansas City, MO	A. W. Petersen	71	88,765	1,250.21	4,500M	Pabst Reburke R A
1949	San Francisco, CA	CA H-F Assn.	78	70,560	904.62	4,500F	Rocky Hill Mont Burke Dusty Jo
1950	Syracuse, NY	R. Austin Backus	69	97,190	1,408.55	7,000M	Smithland Supreme Champion
1951	Des Moines, IA	Iowa H-F Assn.	76	99,460	1,308.68	5,200F	Crescent Beauty Lady Gloria 2d
1952	Roanoke, VA	C. B. Smith	90	117,095	1,301.06	4,200M	Pabst Sir Fobes Burke
1953	St. Paul, MN	Baird & Darcey	76	118,295	1,556.51	7,550M	Weber Haze-Burke Paul
1954	Grand Rapids, MI	C. B. Smith	99	86,625	875.00	3,000F	Sutten Oaks Var Fayne
1955	Harrisburg, PA	R. Austin Backus, Inc.	77	94,490	1,227.14	4,000F	Pabst Roamer Beauty
1956	Milwaukee, WI	Baird & Darcey	73	96,475	1,321.57	9,150F	Zeldenrust Proud Mistress
1957	Indianapolis, IN	Leland Osborn & W. L. Baird	76	88,850	1,169.08	7,050M	Gray View Skyleader
1958	Springfield, MA	W. A. Baker	74	108,400	1,464.86	6,000M	Wis Reward
1959	Omaha, NE	E. A. Dawdy & R. E. Everly	62	95,480	1,540.00	7,200F	Valla Vista Ramona Smoky
1960	Syracuse, NY	Backus, Inc., & Wilcox, Inc.	74	141,022	1,905.71	10,000M	Penstate Pathfinder Ivanhoe
1961	Cleveland, OH	Backus, Inc., & Fenstermaker	79	115,260	1,458.99	7,500M	Skokie Champion

NATIONAL CONVENTION SALES 1962-1984

Year	Place	Sale Manager	No. Head	Total	Ave.	Top	
1962	Roanoke, VA	VA Co-op. Holstein Br. Assn.	75	76,350	1,018.00	3,500F	Wis Sharl
1963	Salt Lake City, UT	Nichols & Thomson	47	90,030	1,915.53	13,100M	Rich-Herd Texal Trademark
1964	Gaithersburg, MD	A. Doty Remsburg	57	111,850	1,962.28	17,000M	Zeldenrust Pontiac
1965	Seattle, WA	M. B. Nichols & A. C. "Whitie" Thomson	43	76,960	1,789.76	7,000F	Piper View Count Johanna Mae
1966	Wichita, KS	M. B. Nichols & A. C. "Whitie" Thomson	37	119,750	3,236.49	25,000M	Fultonway Johann
1967	St. Paul, MN	Piper Bros.	48	134,600	2,804.17	12,000F	Heatherstone Sparkle G
1968	Milwaukee, WI	Holstein-Friesian Assn. of WI	59	163,950	2,802.56	10,000M	Pinehurst Pathfinder
1969	Anaheim, CA	M. B. Nichols	50	219,600	4,392.00	30,000M	Sterk Ivanhoe
1970	Topsfield, MA	W. A. Baker	52	211,650	4,070.19	20,500F	Hilfar Pride Janet
1971	Des Moines, IA	A. C. "Whitie" Thomson & M. B. Nichols	61	268,775	4,406.15	40,000M	Sterk Milly Astro-Twin
1972	Winston-Salem, NC	A. Doty Remsburg & R. Austin Backus, Inc.	68	274,300	4,033.82	32,000M	Duncravin Boehms Countdown
1973	Detroit, MI	Alvin R. Piper & Assoc.	77	372,050	4,831.82	32,000F	Staffordshire Tim Jasmin
1974	Mullica Hill, NJ	R. Austin Backus, Inc.	67	606,400	9,050.75	70,000F	Fran-Lou Pam of Poverty Hollow
1975	Chicago, IL	A. C. "Whitie" Thomson & M. B. Nichols	62	360,500	5,814.52	24,000M	Keystone Illini John
1976	Lancaster, PA	PA Holstein Assn. & R. Austin Backus, Inc.	74	545,100	7,366.22	29,500F	Keystone Joy of Christmas
1977	Columbus, OH	OH Holstein Assn. & R. Austin Backus, Inc.	78	621,000	7,961.54	41,000F	JPG Standout Kandy
1978	Louisville, KY	Alvin R. Piper & Assoc.	88	834,400	9,481.82	131,000M	Paclamar Missile
1979	Denver, CO	David H. Bachmann	126	1,199,800	9,522.22	90,000M	Kanza Pinehurst Burke
1980	Nashville, TN	David H. Bachmann	117	1,224,500	10,465.00	75,000M	Sir C Valor
1981	Baltimore, MD	Remsburg Sale Service	106	1,245,350	11,748.58	70,000M	Fisher-Place Baltimore-ET
1982	Wichita, KS	D. H. Bachmann Co.	123	1,280,300	10,408.94	100,000M	Mr Neutron-ET
1983	Norfolk, VA	VA Holstein Assn. & Backus Assoc.	124	2,214,800	17,861.29	160,000F	Langdonhurst Valiant Nicki
1984	St. Louis, MO	Dallas Burton & Ed Fellers, Jr.	121	2,462,300	20,349.59	255,000F (for 7 head)	
						240,000M	Lylehaven Columbus Mystique-ET

75 Years Of Holstein Sales

Year	No.Sold		Ave.	Year	No. Sold		Ave.
1983*	22,421	65,115,655.00	$2,904.23	1945	11,992	4,173,232.50	348.00
1982	22,729	63,634,180.00	$2,799.69	1944	11,909	3,895,038.75	327.06
1981	22,902	66,109,195.00	2,886.61	1943	9,190	3,154,665.50	343.27
1980	21,099	56,855,401.00	2,694.70	1942	9,178	2,052,838.00	223.62
1979	23,799	55,574,182.50	2,335.15	1941	5,953	1,103,701.25	185.40
1978	20,600	34,522,440.00	1,675.85	1940	5,440	940,523.25	173.25
1977	19,367	23,412,655.00	1,208.89	1939	4,477	681,623.00	152.25
1976	20,808	24,936,752.00	1,198.42	1938	4,082	710,330.00	174.01
1975	14,925	17,022,362.50	1,140.53	1937	5,173	806,433.50	155.89
1974	17,551	19,103,584.00	1,088.46	1936	3,813	624,783.00	163.85
1973	20,415	20,711,000.00	1,014.50	1935	3,439	463,686.00	134.88
1972	20,728	17,238,774.50	831.67	1934	1,950	205,257.00	105.26
1971	15,714	10,975,657.50	698.40	1933	1,563	167,310.50	107.68
1970	16,598	10,885,773.00	655.85	1932	2,023	245,417.50	121.31
1969	17,457	10,227,753.50	585.88	1931	3,689	534,074.50	144.81
1968	17,663	10,233,880.00	579.40	1930	4,637	851,627.00	183.64
1967	19,643	10,358,619.50	527.34	1929	5,086	1,205,901.75	237.10
1966	20,432	10,111,473.00	494.88	1928	5,923	1,229,969.00	207.66
1965	19,785	7,877,945.50	398.18	1927	5,184	1,139,897.50	219.88
1964	16,942	7,071,741.00	417.41	1926	7,466	1,412,960.00	189.25
1963	17,898	7,211,335.00	402.91	1925	8,583	1,245,098.00	145.10
1962	17,379	6,961,557.00	400.57	1924	8,385	1,326,800.00	158.24
1961	16,784	7,199,859.50	428.97	1923	9,422	1,652,673.00	175.41
1960	17,940	7,691,190.50	428.72	1922	8,897	1,896,271.00	213.13
1959	17,657	7,941,481.50	449.76	1921	12,961	3,354,068.00	274.21
1958	18,082	7,795,696.50	431.13	1920	18,836	7,010,214.00	372.12
1957	18,624	6,647,786.50	356.95	1919	18,063	5,852,631.00	324.00
1956	18,990	6,440,866.50	339.17	1918	14,537	3,754,603.50	256.47
1955	18,803	6,225,142.00	331.00	1917	12,728	3,136,544.00	246.43
1954	16,067	5,512,486.00	343.09	1916	12,730	2,474,843.00	194.41
1953	15,791	6,120,440.00	387.59	1915	6,441	1,592,203.00	247.00
1952	16,556	8,013,789.50	484.04	1914	5,832	1,534,806.00	263.00
1951	16,873	8,598,120.00	509.58	1913	4,737	1,178,816.00	248.00
1950	15,155	6,560,060.00	432.87	1912	4,170	814,970.00	195.00
1949	14,759	6,182,073.50	418.87	1911	3,346	685,918.00	205.00
1948	14,875	6,537,406.00	439.49	1910	2,138	396,404.00	185.00
1947	14,926	6,248,681.00	418.64	1909	1,946	329,907.00	170.00
1946	13,610	5,534,787.50	406.67				

SOME ALL-AMERICAN STATISTICS

Listed are the exhibitors that have won All-American honors more than once in the important senior female (2 years and older) classes, with the number of animals and number of awards given.

	No. Of Animals	No. Of Awards
Pabst Farms, Inc., WI	9	10
Carnation Farms, WA	7	10
Hanover Hill (Heffering-Trevena, NY/Ont.	6	8
Tara Hills (Heffering-Houlahan), NY	2	3
Crescent Beauty Farm (Hetts Family), WI	6	8
Maytag Dairy Farms, IA	6	6
Dunloggin, MD	5	7
Chambric Farm, IL	5	5
Mount Victoria Farms, Que.	4	10
Gray View Farms, WI	4	6
Billiwhack Stock Farm, CA	4	4
Paclamar Farms, CO	3	5
Allen Dairy Farms, PA	3	4
Hector I. Astengo (Rosafe), Ont.	3	4
Franlo Farms, MN	3	4
J. W. Innes & Sons, Ont.	3	4
Pinehurst Farms, WI	3	3
Elmwood Farms, IL	3	3
E. H. Ravenscroft, IL	3	3
Dr. H. J. Schmidt (Lavacre), CA	2	6
Raymondale Farm, Que.	2	3
Rocky Hill Farms, CA	2	3
John L. Smith, WA	2	3
Marion Andrew, MD	2	2
Beacon Milling Co., NY	2	2
Bond Haven Farm, Ont.	2	2
Braun's Sunny Lea Farm, WI	2	2
Cornell University, NY	2	2
C.P.R. Supply Farm, Alta	2	2
Dreamstreet Holsteins, NY	2	2
Forsgate Farms, NJ	2	2
Forum Holstein Farm, IA	2	2
C. E. Griffith, OK	2	2
Wm. A. Hayssen (Lakeside), WI	2	2
Tom A. Nunes, Jr., CA	2	2
Oak Ridges Farm, Ont.	2	2

VETERANS OF THE ALL-AMERICAN JUDGES' PANEL

	Years on Panel
Harvey W. Swartz, WI	27
John L. Morris, MD	23
Dr. C. Fred Foreman, IA	22
Eugene W. Nelson, WI	22
Dr. Chas. L. Norton, KS	22
Joe P. Eves, AZ	20
James M. Lewis, OH	20
John W. McKitrick, OH	20
A. C. (Whitie) Thomson, IL	20
Richard E. Keene, NY	19
Elmer N. Hansen, CA	18
Prof. Hilton Boynton, NH	17
Dean H. H. Kildee, IA	17
Henry (Sonny) Bartel, WI	16
Elmer A. Dawdy, KS	16
Allen Hetts, WI	16
Ray Brubacher, Canada	14
Dr. Gordon M. Cairns, MD	14
Paul B. Misner, PA	14
William S. Moscrip, MN	14
Prof. F. W. Atkeson, KS	13
Axel Hansen, MN	13
Elis Knutson, WI	13
Donald V. Seipt, PA	13
Prof. G. E. (Eddie) Gordon, CA	12
Wm. K. Hepburn, Jr., NY	12
Prof. Harold Kaesar, OH	12
C. S. (Dusty) Rhode, IL	12

ALL-TIME ALL-AMERICAN JUDGES PANEL

(Judges with at least three appearances on A-A Panel 1966-1983 plus the Nominating Committee)

J. L. Albright, Henry (Sonny) Bartel, Arlan Bartling, Donald Bay, Robert Beauprez, Howard Binder, C. M. Bottema, Jr., Hilton Boynton, Richard Brooks, Robert Brown, Ray Brubacher, G. M. Cairns, John Carlin, Clyde Chappell, Donald Collins, Glenn Cook, Elmer Dawdy, Dr. David Dickson, Orton Eby, William Etgen, Jimmie Eustace, Doug Fellers, Herbert W. Filk, Fred Foreman, Genie Francisco, Ed Fry, Stu Gibson, Vic Gray, Velmar Green, R. Peter Heffering, David Houck, Merle R. Howard, Vernon Hull, Harold Kaeser, Richard E. Keene, E. A. Keyes, Jerry King, Wally Knapp, Ray Kuehl, Jim Lewis, Lowell Lindsay, Ron Long, Tom Lyon, Doug Maddox, G. B. Marion, Kent Mattson, Richard Mayer, Robert Mayer, Dr. John L. McKitrick, John W. McKitrick, Charles Mickelson, John Morris, Ray Murley, Archie Nelson, Eugene Nelson, Marlowe Nelson, Charles Norton, George Nunes, Tom A. Nunes, Jr., Harry Papageorge, Jimmy Papageorge, Lew Porter, Terry Price, John Rinehart, Harry Roth, Donald Seipt, Dave Smokler, Obie Snider, Bertram Stewart, Ivan Strickler, Robert Strickler, Harvey Swartz, John Tenneson, A. C. "Whitie" Thomson, Harmon Toone, George Trimberger, Willie Watson, Bryce Weiker.

Excellent-97 Holsteins
(as of January 1985)

FEMALES

1. Linden Dictator Wimble Wimpy (3/63)
2. Snowboots Wis Milky Way (7/65)
3. Harborcrest Rose Milly (7/66)
4. Kings Arctic Rose (3/67)
5. Zeldenrust Mistress Roxann (6/67)
6. Johns Lucky Barb (3/68)
7. Future Hope Rocket Flicka (4/71)
8. Gene Acres Felicia May Fury (3/73)
9. Majestic Elms Amy (4/74)
10. Dolvic Fury Bubaleah (6/75)
11. Woodbine Ivanhoe Mollie (1/77)
12. C Glenridge Citation Roxy (8/78)
13. Sully Fobes Bootmaker Dot (1/80)
14. Jan-Com Fond Matt Matilda (2/80)
15. Vernway Echo Violet (11/80)
16. Minco Arlinda Chief Cleo (6/81)
17. Northcroft Ella Elevation (12/81)
18. Vinbett RRD B Fancier (1/82)
19. C Sunnydene Ned Ginger (1/82)
20. Shadowcliff R A Gina (1/82)
21. Brookview Tony Charity (6/84)

MALES

1. Zeldenrust Fond Memory (6/74)
2. C Carlspride Vogel Reflection (9/74)

SURVEY OF STATE HOLSTEIN PUBLICATIONS

Magazine	Circulation	Issues/ Year	Tabloid or Magazine	B & W Page Ad Rate	*Subscription Price	Cost per 1,000 Subscribers
Arizona Sun	3,200	1	M	$255.00	Free	$ 79.69
Arkansas Holstein News	1,200	1	M	100.00	Free	83.33
California Holstein News	1,500	7	M	210.00	$ 5.00	140.00
Colorado Mile High News	1,100	2	M	135.00	Free	122.73
Idaho Holstein Annual	2,400	1	M	150.00	Free	62.50
Illinois Holstein Herald	2,600	4	T	135.00	Free	51.92
		1 Annual	M	170.00	Free	65.38
Iowa Holstein Herald	2,538	5	M	125.00	$ 5.00	49.25
Kansas Holsteins	2,000	2	M	90.00	Free	45.00
Kentucky Holstein News	3,300	3	M	80.00	Free	24.24
Michigan-Indiana Holstein	4,700	11	M	125.00	$ 1.00	26.60
Midwest Holstein News	13,000	8	T	230.00	$ 5.00	17.69
Minnesota Holstein News	2,500	4	M	125.00	$ 5.00	50.00
Mississippi Holstein News	600	2	M	85.00	Free	141.67
Missouri Holstein Journal	4,100	4	M	100.00	$ 5.00	24.39
Nebraska Holstein News	1,554	3	M	85.00	Free	54.70
New England Bulletin	2,553	6	M	182.00	$ 6.00	71.29
New York News	6,845	12	M	245.00	$ 2.00	35.79
Northwest Holstein News	4,600	12	T	210.00	Free	45.65
Ohio Holstein News	2,847	7	M	165.00	$12.00	57.96
Oklahoma Holstein Views	1,450	1	M	100.00	Free	68.97
Oregon Holstein Annual	2,000	1	M	190.00	Free	95.00
Pennsylvania Holstein News	8,041	12	M	245.00	$ 2.00	30.47
South Dakota Holstein News	1,200	3	M	50.00	$ 3.00	41.67
Southeastern Holstein News	3,500	4	M	97.50	$ 5.00	$ 27.86
Tennessee Holstein News	7,050	6	T	240.00	Free	34.04
Texas Holstein News	4,500	3	M	150.00	Free	33.33
Wisconsin Holstein News	7,450	12	M	150.00	$ 5.00	20.13

**Subscription price is for non-members and/or out of state readers.*

Prepared June 1984 by Holstein-Friesian World, Inc.

INTERNATIONAL HERD BOOK SOCIETIES

Asociacion Criadores de Holando Argentino - Argentina
The Holstein-Friesian Association of Australia
Austrian Holstein-Friesian Association
Holstein-Friesian de Belgique - Belgium
Associacao Brasileira de Criadores de Bovinos da Raca Holandesa - Brazil
The Holstein-Friesian Association of Canada
Asociacion de Criadores de Ganado Holandes de Chile - Chile
China Dairy Cattle Association - (Proposed)
Asociacion Colombiana de Holstein-Friesian - Colombia
Asociacion de Criadores de Ganado Holstein de Costa Rica
The Black and White Cattle Association of Denmark
Asociacion Holstein-Friesian del Ecuador - Ecuador
Asociacion de Ganaderos de El Salvador - El Salvador
British Friesian Cattle Society of Great Britain and Ireland
British Holstein Society
Unite Nationale de Selection et de Promotion de la Race Bovine Francaise Frisonne - France
Verband Deutscher Schwarzbruntzuchter e.V - West Germany
Registro Genealogico, Disesepe - Guatemala
National Centre of State Farms - Hungary (Proposed)
Institute of Animal Breeding & Feed Classifying - Hungary (Proposed)
Israel Cattle Breeders' Association, Ltd. - Israel
Associazione Nazionale Allevatori Bovini Della Razza Frisona Italiana - Italy
The Holstein Cattle Association of Japan
The Hokkaido Holstein Association - Japan
Kenya Stud Book - Kenya
Korea Animal Improvement Association (KAIA) - South Korea
de la Federation des Herd Books Luxembourgeois - Luxembourg
Asociacion de Criadores Holstein-Friesian de Mexico - Mexico
Friesch Rundvee-Stamboek-Netherlands
New Zealand Friesian Association - New Zealand
Asociacion de Criadores Holstein Registrado de Peru - Peru
Associacao Portuguesa dos Criadores da Raca Frisia - Portugal
The Friesland Cattle Breeders Association of South Africa
Asociacion Nacional Frisona Espanola - Spain
The Swedish Friesian Cattle Association
The Swedish Holstein Club
Holstein-Friesian Association of America - United States
Sociedad Criadores Holando de Uruguay - Uruguay
Asociacion Venezolana Criadores de Ganado Lechero - Venezuela
Herd book society of Zambia - Zambia

Glossary

ARTIFICIAL INSEMINATION (AI) — the process by which a technician places semen (frozen or liquid) into the mouth of the uterus of a female bovine. Semen is collected from bulls by trained technicians, examined for quality, then diluted with an extender to nourish the sperm and to protect them during the freezing and thawing process. Through these methods, an average sire can produce enough semen for about 40,000 inseminations each year. "Natural service" refers to breeding by the actual bull.

BLUP — "Best Linear Unbiased Predictions". It is a sophisticated, computerized statistical procedure with properties that assure the fewest biases among the current genetic methods now available. This system is used for computing Predicted Differences for Type (PDTs).

BRED HEIFER — a bovine female, usually 18-30 months of age, which has been inseminated and is certified pregnant by an accredited veterinarian. The average length of gestation is 278 days for Holstein females.

BUTTERFAT CONTENT — the percentage of the total volume of milk produced that is fat. Generally, the more milk that is produced, the lower will be the fat content of the milk, since yield of milk is negatively correlated genetically with fat content.

CALF — a bovine less than one year of age.

CERTIFICATE OF REGISTRY — an official document issued by the various breed associations which guarantees the lineage and indentification of all registered animals enrolled in the herd book. A Certificate is issued to every registered animal in the herd book.

CLASSIFICATION SYSTEM — the degree of perfection in type on a scale of 50 to 100 points that an animal has relative to the ideal for that breed. Cattle are classified into one of the following categories: Poor (P), less than 65 points; Fair (F), 65-74; Good (G), 75-79; Good Plus (G+), 80-84; Very Good (VG), 85-89; or Excellent (E), 90+. In addition, there are several descriptive codes which further describe an animal's appearance. All of this information is used for corrective breeding according to guidelines set forth by breed associations. A linear classification system complements this system (see Linear Classification).

COW — a bovine female which has had one or more offspring and is usually two years of age or older. A cow may be in milk (lactating) or dry depending on her breeding status and environmental factors.

Cow Index (CI) — a measure of the transmitting ability of the cow that corresponds to the Predicted Difference indexes for sires.

DUAL PURPOSE — usually breeds of cattle that have been selected for both milk and meat production. Examples are Fleckvieh and Simmental. They almost always produce less milk then breeds selected for intensive milk production (such as U.S. Holsteins).

EMBRYO TRANSFER — the process of flushing one or more embryos from a cow in normal estrus cycle or one that has been induced to superovulate by hormone treatment and transferring the embryo(s) to an estrous synchronized recipient(s) or host. Embryos can be transferred fresh or frozen.

FEED EFFICIENCY — an objective measurement of the ratio of feed consumed to the amount of milk or meat produced from that feed.

GRADE HOLSTEIN — a male or female Holstein whose lineage cannot be identified as registered Holstein and, therefore, cannot possess an official Certificate of Registry, Identified Holstein Report of Pedigree. Most non-registered (grade) Holsteins are essentially purebred, especially since the advent of artificial insemination, since very little breed crossing for dairy purposes has been done in the Holstein breed in recent years.

GOLD MEDAL SIRE — an award recognizing U.S. Holstein bulls that rank in the top 15% of the breed for PDM and PDT, top 25% of the breed for PDF, have repeatabilities of 80% or higher, have PD% Fat not less than -.20, and are free from transmitting undesirable recessive traits.

IDENTIFIED HOLSTEIN FEMALE — a Holstein female whose paternal lineage is guaranteed and recorded in the Holstein Association herd book. The animal is identified by the Association with an Identified Holstein Report which verifies the animal's name, eartag number and birth date and its sire's registered name and herd book number. A sketch or photograph is attached to the Identified Holstein Report to positively identify the animal. (NOTE: The dam is not registered, but may have been positively identified.)

IDENTIFIED HOLSTEIN REPORT — identification form issued by the Holstein Association which guarantees the sire, birth date, color markings and the breeder of a non-registered Holstein female of pure breeding.

LACTATION — the production of milk, initiated at the time of calving and usually continuing 10 to 12 successive months. The length of lactation depends on the time of pregnancy and genetic ability, interacting with nutrition and management factors. In the U.S., lactations are standardized to 305 days for economic reasons and also for appropriate genetic evaluation.

LIFETIME PRODUCTION — the total of milk and butterfat produced during the lifetime of a cow.

LINEAR CLASSIFICATION — a system of type evaluation which measures various biological extremes (traits) in an animal (i.e. levelness of rump and width of rear udder attachment) on a scale of 0 to 50. The method evaluates 14 primary type traits and 14 secondary traits, providing detailed type profiles on each evaluated animal. Combining and analyzing this data also produces exact and detailed genetic predictions for U.S. Holstein sires, allowing for correct mating of cows and bulls. A descriptive type score is also assigned each animal (see Classification System).

OFFICIAL HOLSTEIN PEDIGREE — a document which records the performance of a registered Holstein and the ancestry (paternal and maternal) for as many generations as desired (usually 3 to 5 generations). Official Holstein pedigrees are produced only by the Holstein Association.

PLUS PROVEN SIRE — a bull that has a positive Predicted Difference value for one or more traits. Predicted Difference values can be calculated for nearly every production and type trait.

PREDICTED DIFFERENCE FOR MILK (PDM) — describes the transmitting ability of a sire for milk production. The actual PDM value refers to the amount of milk a given bull's daughters are expected to produce relative to their breed average contemporaries based on the 1982 genetic base. Most other countries use a "breeding value" (twice the PDM value) which represents a sire's total inheritance for milk production. Since a sire transmits only a sample half of his inherited genes for milk production, U.S. dairy farmers use PDM as a measure of a bull's transmitting ability.

PREDICTED DIFFERENCE FOR TYPE (PDT) — measures a sire's transmitting ability for overall type based on the type classifications of his daughters. It can be plus or minus and ranges from about −3.50 to about +3.50 among current U.S. Holstein sires.

PRODUCTION TESTING — the recording of milk weights for each lactating cow for a 24-hour period, one day each month. At each milking (usually two per day), samples of the milk are taken to determine butterfat content (% fat). Production control programs in the U.S. are carried out by state Dairy Herd Improvement Associations (DHIA). About 30% of the more than 11 million dairy cows in the U.S. are enrolled in official production control programs. Herds are enrolled voluntarily by the herd owner.

REGISTERED HOLSTEIN — an animal, male or female, whose lineage has been recorded in the official Holstein Association herd book continuously since the herd book started. This is documented by a Certificate of Registry which verifies the animal's registered name, herd book number and birth date; its sire's registered name and herd book number; its dam's registered name and herd book number; the name of the breeder of the animal; and the name of the last recorded owner. A sketch or photograph is attached to the Certificate giving positive identification of the color markings. The Certificate carries the official seal of the Holstein Association and the signature of the Executive Secretary. About 15 percent of all U.S. Holsteins are registered.

SIRE SUMMARY — all the genetic information available on a bull, including production and conformation transmitting abilities. Sometimes the term is used to refer to the publication of all summaries distributed twice each year by the Holstein Association, the Registered Holstein Total Performance Sire Summary ("Red Book").

TOTAL PERFORMANCE INDEX (TPI) — combines PDM, PD% Fat and PDT in a 3:1:1 ratio to form a value for ranking sires on overall merit. The TPI was developed for screening sires that show the most balance in their transmitting abilities for these economically important traits.

TYPE — a general term referring to the conformation of cow, bull or calf.

Index

ABC Reflection Sovereign, 271-274
  All-American offspring of, 222-223
  Success of in All-Time All-American contest, 229

Advanced Registry testing, 15, 30-31, 51
  affect on Holstein prices, 166-167, 173

Adventure Travel Series, 243

Agricultural Extension Service
  See DHIA programs

Agri-Graphics, Ltd.
  See Holiday ET Extravaganza

Agro Acres Marquis Ned
  All-American offspring of, 223

AI
  See Artificial insemination

Aitken, D. D.
  role of in establishing Holstein-Friesian Association field force, 18
  on Holstein-Friesian Association Reserve Fund, 36

Akins, Zane, 117
  role of in Holstein Sire Development Service, 103
  views on investor money, 195

Albright, J. L.
  See Red & White Holsteins, registration of

Alcartra King Sylvia
  sale of at 1920 National Sale, 170

All-American
  Aged Cows, 224-225
  contest, 218-221
  See also All-Time All-American contest

Allegany-Steuben Holstein Club Sales, 163

Allen Dairy Sale of Champions, 188

Allendairy Glamourous Ivy
  sale of at Pearmont Dispersal, 192

All-Time All-American
  contest, 225-230
  nominating committee, 226

AR testing
  See Advanced Registry testing

Arlinda Holsteins
  See Pawnee Farm Arlinda Chief

Artificial Insemination
  introduction and growth of, 47-49
  affect on sale of bulls, 47-48
  See also Blood typing; PDCA

Association of Breeders of Thoroughbred Holstein Cattle, The, 7

Astengo, Hector I
  See ABC Reflection Sovereign

Auctions, 159-195

Auger, C. J.
  See Winter Place Sale

Augur Sale, Don
  See Don Augur Sale

Averill & Gregory Dispersal, 167

Babcock Test for butterfat, 16

Backus Pedigree Company, 187

Backus, R. Austin, 179
  See also Backus Pedigree Company; Earlville Sale series; Super Duper Sale series

Barney, Fay C.
  See Holstein-Friesian World

Barney, W. B.
  See National Dairy Show, first; Waterloo Dairy Cattle Congress

Barney, W. B. & Associates Home Farm Dispersal, 166

Barns, C. T.
  See Holstein-Friesian Services

Barrett, George
  role of in computerization of Holstein-Friesian Association, 45

Beecher Arlinda Ellen
  high production of, 116-117
  See also Pawnee Farm Arlinda Chief

Bell Farm Rosalind
  sale of in 1932, 174
  See also Wisconsin Admiral Burke Lad

Best Linear Unbiased Prediction
  See BLUP

Bickford, S. L.
  See Osborndale Ivanhoe

Billy Boelyn
  show ring success of, 199

Blood typing, 49-50
  See also artificial insemination; embryo transfer

BLUP, 108

Bond Haven Rag Apple Maple
  All-American offspring of, 223

Bottema, C. M., herd dispersal, 182

Braintrim Elevation Marikin
  See Round Oak Rag Apple Elevation

Breed publications, 235-244
  State publications, 242

Breeders Consignment Sales
  See Promotional sales

Breezewood Patsy Bar Pontiac
  high fat production of, 116

Brentwood Sale series, 171-172

Brood cows
  See Gold Medal Dam award

Brook Bank Holstein herd
  See Wales, Thomas B.

Brookview Tony Charity
  success of in All-American contest, 228

Brooks, W. Richard
  See Paclamar Astronaut; Paclamar Farms; Paclamar Bootmaker

Brothertown herd dispersal, 167
Brown, Charles G.
  See Holstein-Friesian World
Brown, Ira S.
  See Holstein-Friesian World
Brown, Wallace
  See Burkgov Inka De Kol
Burkgov Inka De Kol, 278-282
Cabana Case, 22-23, 170-171
Cabana, Oliver
  See Cabana Case
Caine, George B.
  See Burkgov Inka De Kol; Richmond Black & White Show
Canandaigua sales, 180
Carnation Farms Sale, 187-188
Carnation Governor Prospect
  sale of at 1934 Royal Brentwood National Sale, 174
Carnation Homestead Daisy Madcap
  See Governor of Carnation
Carnation Homestead Inka Mutual
  See Governor of Carnation
Carnation King Sylvia
  sale of in 1918, 168-170
Carnation Milk Farms
  See Carnation King Sylvia; Governor of Carnation; Lakefield Fobes Delight; Sir Inka May
C Carlspride Vogel Reflection
  All-American winnings of, 223-224
Central National Holstein Show, 212-213
Champion Sylvia Johanna
  See Carnation King Sylvia
Chase, Mabel C.
  See Holstein-Friesian World
Chenery, Winthrop W.
  feud with Thomas Whiting, 8
  See also Association of Breeders of Thoroughbred Holstein Cattle, The; Import of Holsteins
Clark, O. G., Classic Sale, 172
Classification, 31-33, 51-52
  See also Descriptive Classification; Linear Classification; SET program
Cleveland, S.D.W., 167
Cloninger, Homer
  See Penstate Ivanhoe Star
Clothilde
  high butterfat production of, 12
Club sales
  See Promotional sales
C Norton Court Model Vee
  success of produce in All-Time All-American contest, 229
Code of show ring ethics
  See Holstein shows, code of ethics
Cole, Charlie, 22
  See also Cabana Case
Color controversy
  See Red & White Holsteins
Conformation
  See Classification; Descriptive Classification; True Type program
Consignment sales
  See Promotional sales; Southfork Sale
Contract terms
  See Sale conditions and terms
Cow Testing Association
  See DHIA programs; Herd testing programs
Crescent Beauty-Admirals Sale, 179, 188
Curtiss Breeding Service
  sale of bulls from, 192
  See also Pabst Sir Roburke Rag Apple; Pawnee Farm Arlinda Chief; Romandale Reflection Marquis
Dairy Cattle Congress
  See Waterloo Dairy Cattle Congress
Dairy Herd Improvement Association
  See DHIA
Dairy Shrine, 246-248
Davidson, W. G.
  See Brentwood Sale series
Dawdy, Elmer
  See Ideal Fury Reflector; Wis Burke Ideal
Dawdy, Max
  See Penstate Ivanhoe Star
Dawson, Florence
  Association service of, 117
Descriptive Classification, 67-71, 106-107
Designer Fashion Sale
  1975, 188
  1981, 192
  1983, 193
Detch, Charles
  See Sale ethics
DHIA programs, 51, 73-74
DHIR testing program, 73-74
Diamond J Holstein Dispersal, 186
Dispersal sales, 164-165
Distinguished Junior Member program
  See Youth programs
Dolan, Mike and Marion
  See Drifty Farm Dispersal
Don Augur Sale, 181, 184
Dowager No. 7
  high milk production of, 7
Dreamstreet Holsteins, 1982 sale, 193
  See also Pammies Citation Paula-ET
Dreamstreet Sexation Diligent
  sale of at 1983 Designer Fashion Sale, 193
Drifty Farms Dispersal, 189
Dunloggin Dispersal, 176-177, 265
Dunloggin Duchess Anna
  sale of at 1941 Royal Brentwood, 175

Dunloggin French Mistress
  See Dunloggin Woodmaster
Dunloggin Happy Mistress
  See Dunloggin Woodmaster
Dunloggin Strath Var
  sale of at 1951 Sutten Oaks
    Dispersal, 179
Dunloggin Tidy Miss
  sale of at 1941 Royal Brentwood, 175
Dunloggin Woodmaster, 263-266
  See also Dunloggin Dispersal
duPont, Henry F.
  See Winterthur herd dispersal
Dutch Friesian Association of
  America, 8-10
Earlville Sale series, 171
  See also Super Duper Sale series
Eastern National Holstein Show, 211-212
Elmwood partial dispersal, 174
Embryo transfer, 126-128
  See also Holiday ET Extravaganza; Southfork Sale
Erickson, John
  herd dispersal, 171
  See also Wisconsin Admiral Burke Lad
ET
  See Embryo transfer
Eves, Joe P.
  See Dairy Shrine
Export of Holsteins, 79, 80-97, 128-132
  See also Burkgov Inka De Kol; Carnation Farms
    Sale; Embryo transfer; Foreign Agricultural
    Service Cooperator Agreement;
    Holstein-Friesian Services; Pabst Ideal;
    Sale ethics
Extension program
  See Field services
Felicia May Syndicate
  See Gene-Acres Felicia May, sale of
FFA
  See Youth programs
Field services, 17-19
  expansion of, 38-40, 53-55
  fee-for-service system, 71-72
  states' acceptance of national
    program, 104-105
  See also Classification; Genetic Evaluation
    & Management Service; Holstein Sire
    Development Service; State & local Holstein
    associations; Youth programs
Fishler, Lester, Dispersal
  See Pawnee Farm Arlinda Chief
Food For Freedom program, 37
Foreign Agricultural Services Cooperator
  Agreement, 6-point plan, 82
Foster, O. F.
  See Lakefield Farms Dispersal
4-H
  See Youth programs
Fraudulent practices
  See Cabana Case; Cole, Charlie; Rearick Case
French, Colonel G. Watson
  on AR testing, 30
Fuhriman, Wendall
  See Burkgov Inka De Kol
Future Hope Reflector Blacky
  sale of at Tara Hills Dispersal, 185
  See also Pammies Citation Paula-ET
Gardner, Malcolm H., 16-17
Gene-Acres Felicia May Fury
  success of in All-Time All-American contest, 227
  sale of at Crescent Beauty-Admirals Sale, 188
  See also Ideal Fury Reflector
General Cochran Farm Dispersal, 179
Genetic Evaluation & Management program
    (GEM), 100-101
Glendell Arlinda Chief, 324-327
Gold Medal Dam award, 55
Gold Medal Sire award, 33
Governor of Carnation, 259-263
Grade Holsteins
  See Identified grade Holsteins
Gray View BD Crissy
  All-American offspring of, 224
Gray View Coral Shamrock
  sale of at World Premiere Sale, 186
Gray View Crisscross
  All-American winning of and All-American
    offspring of, 224
  success of in All-Time All-American contest, 229
Groff, Earl
  See Osborndale Ivanhoe
Guarantees of performance
  See Sale terms and conditions
Hall of Fame Sale, 187
Hanover Hill Sales & Service
  See Designer Fashion Sale
Hanover Hill Sale, 1972, 187
Hanover Hill Sheik Barb-ET
  sale of at Odyssey Elite Sale, 193
Hanover-Hill Triple Threat-Red
  sale of at 1972 Hanover Hill Sale, 187
Hansen, Axel
  See True Type program
Harborcrest Rose Milly
  success of in All-Time All-American contest, 227
  See also All-American Aged Cows;
    Paclamar Astronaut
Hargrove & Arnold
  See Triune Papoose Piebe
Harrisburg Association, 63-64
Hartman, Ken and JoAnn
  See Glendell Arlinda Chief
Hartsook, Fred
  and outbreak of hoof and mouth disease, 171
Hastings, Eugene M.
  See Holstein-Friesian World
Hastings, Joel P.
  See Holstein-Friesian World
Hastings, Robert H.
  See Holstein-Friesian World

Hazelwood Holstein Farm
show ring success of, 201
Health, animal
guarantees on sale animals, 167-168
See also Hoof and mouth disease; Import of Holsteins; PDCA; Undesirable recessives
Heffering, R. Peter
See All-American Aged Cows
Heisner. E. E.
See Blood typing
Herd Builder program, 101
Herdmate comparison as method of evaluation, 75
Herd testing programs, 16-17; 30-31
See also HIR; DHIA; DHIR
Hetts, Allen
See Central National Holstein Show; Crescent Beauty-Admirals Sale
HFS
See Holstein-Friesian Services, Inc.
Hill, James F.
See Holstein-Friesian World
Hilltop Hanover Sale, 1980, 192
HIR testing, 30-31, 51
Holiday ET Extravaganza, 191
Holstein-Friesian Association finances of, 36
formation of, 9-10
managerial structure, 15, 41-43, 119-124
office & structural changes at, 13-14, 35, 45-46, 59, 61-62
role of in show policies, 217
Services of
See Field services; Holstein-Friesian Services; Holstein Sire Development Service
Holstein-Friesian Register, 235
Holstein-Friesian Registry Association Inc.
See Harrisburg Association
Holstein-Friesian Services Inc., 84-92
role of in export of ETs, 127
Holstein-Friesian World, 235-244
contributions to integrity in livestock photography, 240-241
relationship with Holstein-Friesian Association, 243
"Holstein Holsum Milk"
See Promotional efforts of Holstein-Friesian Association
Holstein Identification System (HIS)
See Identified grade Holsteins
Holstein sales
See Auctions: Dispersal sales; Promotional sales
Holstein Sire Development Service, 102-104
Holstein shows, 197-232
code of ethics for, 63
contributions to development of the breed, 198-199, 230-231
local, state, regional shows, 215-217
Honor List, 243
Honor List sires
See Penstate Ivanhoe Star
Hoof and mouth disease
affect on imports, 8
affect on public sales, 167, 168, 171
outbreak halts national show, 201
See also health, animal
Hope, Ronald
See Round Oak Rag Apple Elevation
Houghton, Charles
See Association of Breeders of Thoroughbred Holstein Cattle, The
Houghton, Frederick, L., 12-14-29
See also Holstein-Friesian Register; Proxy battle; True Type program
Houlahan, James J.
See All-American Aged Cows
Householder, Glen M.
See Wisconsin Admiral Burke Lad
Hoxie, Solomon
influence on herd classification program, 31-32
role in production testing, 16
See also Advanced Registry testing; Dutch Friesian Association; Holstein-Friesian Association, formation of
HSDS
See Holstein Sire Development Service
Huidekoper, Edgar
See Billy Boelyn, show ring success of
Hungary Program Development, 129-132
Ideal Fury Reflector, 298-301
All-American offspring of, 223
success of in All-Time All-American contest, 229
Identified grade Holsteins, 111-112
Import of Holsteins, 6-9, 175-176
See also Mount Victoria Dispersal
International Dairy Show, 204-206
International Operating Agreement, 129
International marketing
See Export of Holsteins; Holstein-Friesian Services
Investor groups, 190-191, 194-195
affect on sale prices, 186-187
See also Felicia May Syndicate; Leadfield Associates; Southfork Sale
Jarvis, William
See Import of Holsteins
Jayne, Phil
See Glendell Arlinda Chief
Johanna Rag Apple Pabst, 256-259
sale of in 1926 Clark Classic, 172, 258
Johnson, Edward
See Diamond J Holstein Dispersal
JPG Standout Kandy
See All-American Aged Cows
Junior Holstein members
See Youth programs
Kanza Matt Tippy
See No-Na-Me Fond Matt
Kellogg, O. U.
See Cabana Case
Kellogg, Mrs. Waldo
See Osborndale Ivanhoe
Kessen, Robert B.
Association service of, 117
Kildee, H. H.
See Dairy Shrine

King Bessie Senator
All-American winnings of, 221
sale of at 1938 Elmwood dispersal, 175
King Segis Pontiac Alcartra
sale of in 1911, 167
Kliewer, Ray
See Holstein-Friesian Services, Inc.
Kriemheld Dispersal
See Miller, Gerrit S., herd dispersal
Lakefield Farms Dispersal, 180
Lakefield Fobes Delight
sale of at Lakefield Farms Dispersal, 180
Lakefield Fond Delight Fobes
sale of at Lessard Dispersal, 182
Larson, Charles J.
Association service of, 114-116
Lashbrook, A. J.
herd dispersal, 180
Lashbrook Pearl Ormsby
sale of at Royal Brentwood Sale, 174
See also Dunloggin Woodmaster;
Lashbrook, A. J., herd dispersal
Leadfield Associates
See Md-Maple-Lawn Marquis Glamour
LeRoy, Herman
See Import of Holsteins
Lessard Dispersal, 182
Lilith Pauline De Kol
high butterfat production of, 166
Lincklaen, John
See Import of Holsteins
Linden Dictator Wimble Wimpy
sale of at Sunny Lea Dispersal, 182
See also All-American Aged Cows
Lindskoog, Wallace N.
See Pawnee Farm Arlinda Chief
Linear Classification, 125
Linker, Philip
See Johanna Rag Apple Pabst
Long-Haven Wayne Stephanie
sale of at Southfork Sale, 194
Lowden, Frank O.
election of as President of Holstein-Friesian
Association, 25
McCauley, T. B.
See Mount Victoria Farms
Madison Square Garden dairy cattle show
of 1887, 10-12
Mallary Farm
See Montvic Rag Apple Colantha Abbekerk
Man-O-War
All-American offspring of, 222
Maple View Dispersal, 182
Marketing of Holsteins
See Auctions; Export of Holsteins;
Holstein-Friesian Services, Inc.
Mathiesen, H. A.
See Progressive Breeders Registry program
Maytag Dairy Farms
All-American winnings of, 222
May Walker Ollie Homestead
sale of at Minnesota Holstein Company
Dispersal, 173
McDowell, Irvin
See Ideal Fury Reflector
McKown, Robert M.
See Holstein-Friesian World
McVay, Hobart and Marion
See No-Na-Me Fond Matt
Md-Maple-Lawn Marquis Glamour
sale of in Allen Dairy Sale of Champions, 188
Mercedes
high butterfat production of, 12
Merchandising registered Holsteins
See Auctions; Dispersal sales; Promotional
sales; Sale ethics
MGM Grand Sale, 189
Milk constituents, 38
See also Solids-not-fat testing
Milk marketing
See Promotional efforts of Holstein-Friesian
Association
Miller, George
See Round Oak Rag Apple Elevation
Miller, Gerrit S.
herd dispersal, 174
See also Billy Boelyn, show ring success of;
Import of Holsteins
Minnesota Holstein Company Dispersal, 172-173
Minnow Creek Eden Delight
high production of, 59
sale of at Lakefield Farms Dispersal, 180
Misner, Paul B.
See Brentwood Sale series; Dunloggin Dispersal;
Dunloggin Woodmaster; *Holstein-Friesian
World*
Miss Milliehoe
sale of at Winter Place Sale, 189
Mix, Maurice
on role of classifiers, 68
Montvic Bonheur Pietje B
sale of at Mount Victoria Dispersal, 176
sale of at 1944 Royal Brentwood Sale, 177-178
Montvic Lochinvar
sale of at Mount Victoria Dispersal, 176
Montvic Rag Apple Colantha Abbekerk
sale of at Mount Victoria Dispersal, 176
See also Johanna Rag Apple Pabst
Montvic Renown
sale of to Curtiss Candy Co., 176
Mooseheart Farm
See Pabst Sir Roburke Rag Apple
Morris, A. W. & Sons
consignors of high selling bulls at
1920 national sale, 170
Morrow, David L.
See Holstein-Friesian World
Moscrip, W. S.
role of in All-American program, 219
See also Classification; True Type program
Mount Victoria Farms
Dispersal, 176
purchase of Johanna Rag Apple Pabst, 172
See also All-American Aged Cows;
Johanna Rag Apple Pabst

Mowry Prince Corrine
high milk production of, 116
Moyer, H. A.
national sale series of, 168
National All-American Sale, 182
National Dairy Cattle Congress
See Waterloo Dairy Cattle Congress
National Dairy Show, first, 200
National Holstein Futurity, 208
National show concept, three, 206-209
Natwick, J.
See Dunloggin Dispersal
NEBA (Northeastern Breeders Association)
See Whirlhill Kingpin
Nelson, Richard
See Blood typing
Nesbitt, Arthur
role of in Pennsylvania field force, 105
Netherland Prince
show ring success of, 199
New Hampshire, University of
See Progressive Breeders Registry award
Nichols, M. B.
See Progressive Breeders Registry award; Romandale Reflection Marquis
No-Na-Me Fond Matt, 301-305
Northcroft Ella Elevation
success of in All-Time All-American contest, 227, 228
See also All-American Aged Cows; Round Oak Rag Apple Elevation
North Star Gelschecola Champion
first Gold Medal Sire, 34
Norton, Horace W. (Hod), 46
on "Food For Freedom" program, 37
on need for reorganization of Holstein-Friesian Association, 41
See also PDCA
Nunes, Tom A. Jr.
See Western National Holstein Show
Nutter, Fred
See Maple View Herd Dispersal
Odyssey Elite Sale, 193
Off-color registration, 64-67
Osborn, James
See Osborndale Ivanhoe; Whirlhill Kingpin
Osborndale Farms
Dispersal, 180-181
purchase of Sir Bess Ormsby May, 173
See also Dunloggin Tidy Miss; Osborndale Ivanhoe
Osborndale Gay Fobes
See Whirlhill Kingpin
Osborndale Ivanhoe, 285-289
All-American offspring of, 223
Pabst Comet
See Wisconsin Admiral Burke Lad, All-American offspring of
Pabst Farms
All-American winnings of, 222
Dispersal, 182-183
See also Pabst Sir Roburke Rag Apple; Wisconsin Admiral Burke Lad
Pabst, Fred
See True Type program
Pabst Ideal
importance of in export efforts, 83, 183
Pabst Korndyke Cornflower
sale of at 1920 national sale, 170
Pabst Regal
See Wisconsin Admiral Burke Lad, All-American offspring of
Pabst Roamer
See Wisconsin Admiral Burke Lad, All-American offspring of
Pabst Sir Roburke Rag Apple, 276-278
Paclamar Astronaut, 316-320
sale of in 1964 national sale, 184
Paclamar Bootmaker, 313-315
sale of in Paclamar Farms Sale, 185
Paclamar Farms
export of herd to Italy, 90
Sale, 185
See also All-American Aged Cows; Paclamar Astronaut
Paclamar Tidal Wave
sale of at Southfork Sale, 194
Pammies Citation Paula-ET
sale of at 1980 Hilltop Hanover Sale, 192
Panciera, Aldo
See Osborndale Ivanhoe
Patterson, F. W.
See Holstein-Friesian Association, formation of
Pawnee Farm Arlinda Chief, 306-309
See also Pawnee Farm Glenvue Beauty
Pawnee Farm Dispersal, 182
Pawnee Farm Glenvue Beauty
sale of at Pawnee Farm Dispersal, 182
See also Pawnee Farm Arlinda Chief
PBR
See Progressive Breeder Registry award
PDCA, 244-245
role of in standardizing of production testing, 73
Pearmont Dispersal, 192
Pennsylvania All-American Dairy Show
See Eastern National Holstein Show
Penstate Ivanhoe Star, 309-312
Penstate Lucifer Anna Star
See Penstate Ivanhoe Star
Performance Qualified shows
See Production Qualified shows
Perry, Enos J.
introduces artificial insemination to U.S. dairymen, 47
Photography, livestock, 240-242
Piek, Joseph E.
See Johanna Rag Apple Pabst
Plowman, Dean
and predicted difference method of sire evaluation, 75
Polinder, Henry
See Western National Holstein Show
Pound, James F.
on Genetic Evaluation & Management Service, 100-101
on three national show concept, 206

Prescott, Maurice S., 238
role of in All-American program, 219
See also Holstein-Friesian World
Prescott, W. Theodore
See Holstein-Friesian World
Prescott, William A.
See Holstein-Friesian World
President's "E" Award, 90
Prestige of Lakehurst
All-American offspring of, 223
Price, Frank T.
See Holstein-Friesian World
Production Qualified shows, 209-210
Production testing, 50, 72-74
See also Advanced Registry testing; Babcock Test for butterfat; DHIA; DHIR; Herd testing programs; HIR; PDCA
Progeny testing
See Sire evaluation
Progressive Breeders Registry award, 34-35
Promotional efforts of Holstein-Friesian Association, 36, 94, 121
Promotional sales, 163-164, 178
Proxy battle, 12-13, 23-25
Publications
See Breed publications
Purebred Dairy Cattle Association
See PDCA
Quality Fobes Abbekerk Gay
See Osborndale Ivanhoe
Rasmussen, Robert V.
See Elmwood partial dispersal; King Bessie Senator
Rath, Gunther
See Pabst Ideal
R. Austin Backus Inc.
See Backus Pedigree Co.
Ravenglen Farms
See King Bessie Senator
Rearick Case, 112-113
Recessive traits
See Undesirable recessives
Red & White Holsteins
registration of, 64-67, 111-112
Remick, Jerome H.
See Dunloggin Woodmaster
Richmond Black & White Show, 215-216
Rocky Hill Mont Burke Dusty Jo
See All-American Aged Cows
Romandale Farms
See Romandale Reflection Marquis
Romandale Reflection Marquis, 289-291
All-American offspring of, 222
Rosafe Farm
See ABC Reflection Sovereign
Rosehill Fayne Wayne
See All-American Aged Cows
Round Oak Ivanhoe Eve
See Osborndale Ivanhoe; Round Oak Rag Apple Elevation
Round Oak Rag Apple Elevation, 320-324
Royal Erinwood Sale, 188
Rumler, Robert H.
election of as Assistant Executive Secretary, 43, 46
as Executive Chairman of Holstein-Friesian Association, 117
role of in export effort, 81, 83
at dedication of Dairy Shrine, 246-247
on genetic progress, 60
Sadie Vale Concordia
high butterfat production of, 167
Sale ethics, 128-129, 193-194
Sale terms and conditions, 184-185
See also Don Augur Sale; Sir Bess Ormsby Burke Fobes; Southfork Sale; Winter Place Sale
Sandberg, Archie
See Wisconsin Admiral Burke Lad
Schmidt, Harold J.
on sire evaluation, 75
Schneider Brothers
See No-Na-Me Fond Matt
Sears, James L.
See Searsfarm Dean Ada Imperial
Searsfarm Dean Ada Imperial, 282-284
Seaverns, Houghton, 29
Selkirk, Robert
See Western National Holstein Show
SET program, 109
See also Descriptive Classification
Sexton, Jack
See Paclamar Bootmaker
Shadowcliff RA Gina
success of in All-Time All-American contest, 228
Shaw, Harold J.
See Shaw's Dauntless-Dunloggin Sale
Shaw's Dauntless-Dunloggin Sale, 179
Show Ring Ethics, Code of
See Holstein shows, code of ethics
Silver Medal Sire
See Sire recognition programs
Sir Bess Ormsby Burke Fobes
sale of in Osborndale Dispersal, 180
Sir Bess Ormsby May
sale of at 1927 Minnesota Holstein Company Dispersal, 172
Sire evaluation, 74-76, 107-109
See also Holstein Sire Development Service; SET program; Total Performance Index
Sire recognition programs, 33-34
Sir Inka May
sale of in 1925 Brentwood Sale, 171
Skagvale Graceful Hattie
high milk production of, 113
Sluiter Brothers
legal battle with Holstein-Friesian Association, 22
Smiths & Powell Co.
show ring success of Netherland Prince, 199
See also Clothilde
Snowboots Wis Milky Way
See Paclamar Bootmaker
Solids-not-fat testing, 73
Southfork Sale, 194

Sovereign Cochran
sale of in General Cochran Farm Dispersal, 179
State and local Holstein
associations, 52-53
relationship with Holstein-Friesian Association, 104-106
See also Field services; Holstein shows, local, state and regional
Stauffer, Clarence
success in Genetic Evaluation & Management Services, 100
Stebbins, Robert
Association service of, 117
Stevens, Ralph J.
See Holstein-Friesian World
Stevens, Ward W.
See Holstein-Friesian World
Straight-Pine Elevation Pete
participation in Holstein Sire Development Service program, 103
Stroh, Charles
See Osborndale Ivanhoe; Whirlhill Kingpin
Strohmeyer, Harry A. Jr., 241
Sunny Knolls Invitational, 188
Sunny Lea Dispersal, 182
Super Duper Sale series, 179
Superiority of Holstein breed, 7, 10-12, 37, 94-95, 129-132
Sutten Oaks Dispersal, 179
Tara Hills Dispersal, 185
Tara-Hills Pride Lucky Barb
sale at 1972 Hanover Hill Sale, 187
Thornlea Texal Supreme
All-American offspring of, 223
Total Performance Index, 107-108
Trevena, Kenneth
See All-American Aged Cows
Trimberger, George
role of in Descriptive Classification program, 69
on balance of production and type, 230-231
Triune Papoose Piebe
All-American winnings of, 221
show ring success of, 200
True Type program, 25-27, 109-111
Undesirable recessives, 113-114
Uniform Rules For Official Testing
See PDCA
Utah State Industrial School
See Wisconsin Admiral Burke Lad
Wales, Thomas B., 12-13
herd dispersal of, 165-166
See also Holstein-Friesian Association, formation of; Mercedes; Proxy battle
Wapa Bootmaker Mandy
success of in All-Time All-American contest, 228
Waterloo Dairy Cattle Congress, 203-204
Weber Burke Dusty Jo
sale of at 1941 Western Classic, 175
See also Wisconsin Admiral Burke Lad
Weber Hazelwood Burke Raven
See Wisconsin Admiral Burke Lad, All-American offspring of
Western International Dairy Exposition
See Western National Holstein Show
Western National Holstein Show, 213-214
Whirlhill Kingpin, 294-297
sale of at Whirlwind Hill Dispersal, 181
Whirlwind Hill Dispersal, 181
Whiting, Thomas E.
See Import of Holsteins; Chenery, Winthrop W., feud with Thomas Whiting
W.I.D.E.
See Western National Holstein Show
Winter Place Sale, 189
Winterthur herd dispersal, 185
Wis Burke Ideal, 274-276
Wisconsin Admiral Burke Lad, 266-269
All-American offspring of, 222
Wisconsin Showcase Sale series, 182
Wisconsin State Reformatory
See Wisconsin Admiral Burke Lad
Wis Leader
See Wisconsin Admiral Burke Lad, All-American offspring of
Wis Maestro
sale of at Osborndale Dispersal, 180
World Dairy Expo
See Central National Holstein Show
World Friesian Conferences, 128
World Premiere Sale, 186
See also Central National Holstein Show
Yeomans, Theron G.
See Holstein-Friesian Association, formation of
Youth programs, 20-21
See also Central National Holstein Show; Eastern National Holstein Show
Zeldenrust Fond Memory
success of in All-Time All-American contest, 229
All-American winnings of, 223-224
Zeldenrust Pontiac Korndyke
high lifetime production of, 59